JÚPITER FIGUERA   JUAN GUERRERO

DESIGN OF ELECTRICAL

INSTALLATION IN HOMES

Master the Basics of Residential
Electrical Installations

DESIGN OF ELECTRICAL INSTALLATIONS IN HOMES
Júpiter Figuera Yibirín
Juan Guerrero Márquez

Copyright © 2022 by Júpiter Figuera Yibirin and Juan Guerrero Márquez

Peer review:
Prof. Carlos Lezama
Prof. Ernesto Leal

Cover, layout, drawings and illustrations:
Prof. Júpiter Figuera Yibirín

ISBN:

First English edition
Buenos Aires, Argentina, December 2022

This book is dedicated to:

My always beautiful wife Isabel
My grandchildren Celeste, Brisa, Karlene, Karla, Katherine and Mateo.
The Argentinean people who welcomed me during a difficult stage of my life
(Júpiter Figuera Yibirín)

My dear wife, Alba
My beloved children Ninel, Juan, Libia and Miriam
(Juan Guerrero Márquez)

Acknowledgments:

We appreciate the exhaustive review of the originals of this work by Prof. Carlos Lezama; although unfinished by the chance of fate, his laborious task will always accompany us. We also acknowledge the careful review carried out by the writer Ernesto Leal to perfect the final style.

INTRODUCCIÓN

TO WHOM THIS BOOK IS ADDRESSED

This book is aimed at those who design electrical installations. It deals with the study of outlets and switches used in residential, commercial and industrial electrical installations. With regard to outlets, its characteristics, capacity and wiring are studied as well as the relationship of the different types of outlets with the electrical safety standards. The ground fault circuit interrupter (GFCI) is studied, as well as its operation and wiring, limitations and types. Multiconductor circuits are a subject of this chapter. The arc fault interrupter is mentioned. Also, starting from basic concepts, the different types of circuit breakers (SPST, SPDT) are presented, describing their operation and their use in electrical installations. Pictorial diagrams are shown of the wiring used to turn on luminaries from different places in a residence or building through the use of simple (single-pole), three-way and four-way switches. Similarly, it establishes how outlets, switches and lighting fixtures should be placed in the rooms of a residence. The most common electrical equipment and appliances in a residential unit are mentioned and indications are given on the distances that the outlets should maintain among themselves and with respect to the furniture found in the different spaces of a dwelling. The study takes into account both indoor and outdoor outlets.

REQUIRED KNOWLEDGE

To approach the study of this book, the basic elements of electricity and fundamental arithmetic operations are required.

REGULATIONS ON WHICH THIS BOOK IS BASED

Throughout the text the adherence of the content to the rules that govern the calculation of residential electrical installations is mentioned. For this, regulations of Latin American countries and the National Electrical Code (NEC) of the United States were consulted. The latter has served as the basis for the drafting of the codes of several Latin American countries. In the case of Venezuela, the National Electrical Code has been translated into Spanish. This code was adapted to Venezuelan characteristics and designated as the National Electrical Code (CEN), because "the construction procedures and materials used in Venezuela are the same in both countries."

ILLUSTRATIONS

In the development of the content abundant use is made of figures related to theoretical concepts. In this way it seeks to achieve a greater understanding of the study material. The transformation of the exposed ideas into beautiful illustrations, very close to what reality presents, makes reading the text more attractive and complements the apprehension of knowledge.

ORGANIZATION

The book is divided into sections numbered in consecutive order. An appreciable number of examples is presented and at the end of it theoretical questions and problems are proposed in relation to what has been studied. Numerous illustrations, referring to the topics described, reinforce the theoretical approaches. The tables included provide data and particularities of the conduits and boxes that characterize electrical installations and are useful for selecting them. These tables were adapted, for the most part, from the electrical codes that govern the design of the electrical installations.

CONTENTS

TOMACORRIENTES

1. CHARACTERISTICS OF THE OUTLETS

In any residence, office, business, industry or place where human activities are carried out, it is safe to find electrical outlets. Such electrical components are defined as contact devices to which a plug is connected in order to supply power to appliances and equipment that use electricity. In other words, outlets are used to supply electrical power to a wide variety of electrical or electronic appliances, including fans, computers, refrigerators, washing machines, microwaves, radios, TVs, stereos and air conditioners. Although visible in an electrical installation, electrical outlets have inside them characteristics that a specialist in the electrical area should know. That is why it is important to study their parts and how they should be connected to the conductors that supply power to them.

There is a great variety of outlets whose construction depends on their intended use. Basically, when it comes to residential use, we can find single outlets and double outlets. The former are used to connect a single piece of equipment or appliance, while double outlets can be used to connect two pieces of equipment. Nowadays it is not common to use single outlets along an electrical installation; double outlets are preferred because of their higher connection capacity, except in those specialized applications that require their use. Such would be the case of outlets for equipment such as air conditioners, dryers, electric pumps, electric stoves and water heaters.

To make use of electrical energy, plugs are used whose shapes adapt to the geometry of the outlets in order to be able to penetrate them. Basically, the front and rear parts of the outlets are insulating and made of a plastic or nylon material, while their internal parts are metallic and from them emerge the contact points for the conductors that are connected to them.

Fig. 1 shows an outlet that is frequently found in our homes. It is a device with two plates, to which terminal screws are attached for the connection of the phase and neutral conductors. In this type of *non-polarized* outlet there is no grounding terminal and its terminals can receive, in an interchangeable way, the phase and neutral conductors. Also, the slots must be connected to a plug that has two pins of equal dimensions. The presence of this type of outlet in residential electrical installations is a latent danger for people, because it does not offer protection in the event that the armatures of the connected equipment come into contact with the phase conductor. The risk of electrocution is high and it is necessary to change the accentuated habit of placing this type of outlet. The term *non-polarized refers to the fact that the hot and neutral conductors can be connected to any of the receptacle's screws.* The lack of polarization introduces an additional factor of insecurity that we will discuss later.

Another alternative that can be found is illustrated in **Fig. 2**. It is a polarized outlet, but without a ground terminal. No attempt should be made to mechanically connect the phase conductor to the neutral or vice versa. This alternative adds a certain security to the installation, but does not avoid the problem of possible electrocution derived from the lack of the grounding terminal. The difference in the sizes of the slots and the dimensions of the plug pins that bring the equipment to be connected guarantee that an inversion between the phase and neutral terminals will not take place, unless, mechanically, it is forced to make the mistake of producing the inversion phase–neutral. The hot conductor must be connected to the tan terminal, and the neutral must be connected to the silver terminal.

A safer outlet than the previous ones is shown in **Fig. 3**. Its use in electrical installations considerably reduces the risk of electrocution. As can be seen, the outlett has two slots of different sizes: *the small one corresponds to the phase and the larger one to the neutral.* In addition, there is a semicircular hole to which the grounding terminal of the device is connected. It is therefore a *double polarized socket with a grounding terminal.* The latter is connected, by means of the plug, to the ground of the equipment to be used. This reduces the risk to the person using electricity when connecting a domestic or industrial appliance.

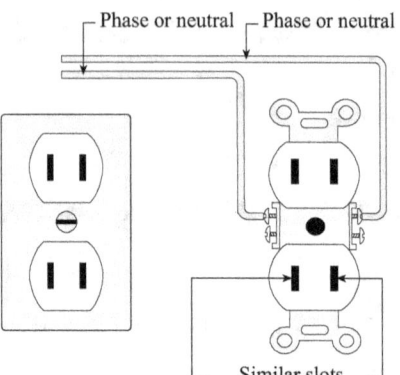

Fig. 1 Double outlet, non-polarized and without grounding terminal. The phase and neutral slots are of equal size.

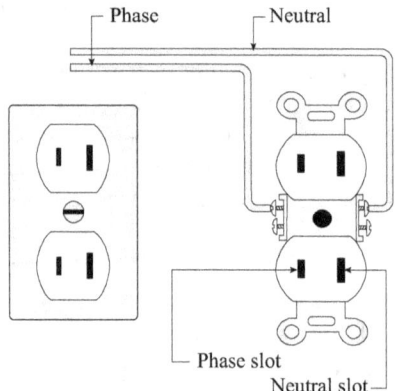

Fig. 2 Double outlet, polarized, without grounding terminal. The phase slot is smaller than the neutral slot.

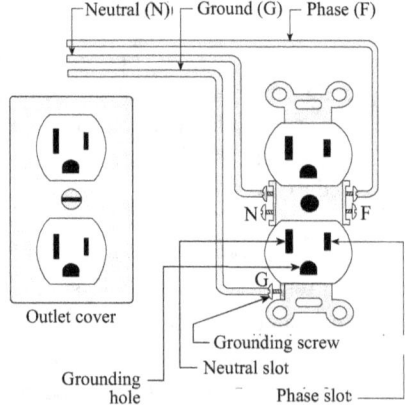

Fig. 3 Double socket, polarized, with grounding terminal. Phase and neutral slots have different sizes. A corresponding grounding hole is added.

Not all outlets are of the same quality. Poor quality ones are prone to cracking, shorting, and creating poor connections that can cause fires. The medium quality ones can be used in corridors and rooms while in the kitchen, the laundry room and other places where their use is frequent, good quality outlets should be used. An indication of the good quality of the outlets is found in the marks printed on their front face. The voltage and current for which it was designed are indicated there (see **Fig. 4**), according to the standards of the NEMA (*National Electrical Manufacturers Association*) and the UL standard (*Underwriters Laboratories*), which indicates the subjection to rigorous tests of the device. and its approval as a secure item.

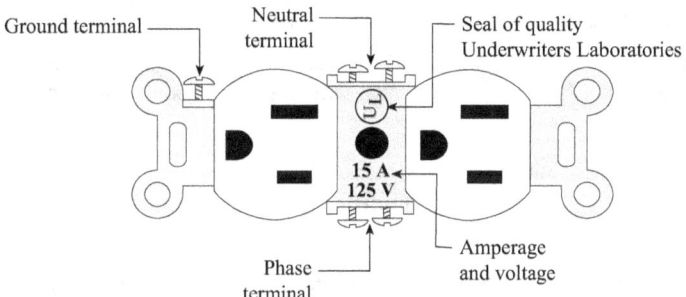

Fig. 4 The markings on amperage and voltage must be present on an outlet. The UL seal guarantees its quality.

Outlets are classified according to the voltage and current they support. Thus we find 125 V/15 A outlets, which must be used on 120 volt branch circuits, and the current of the equipment connected to them must not exceed 15 amps, even when connected to 20 A branch circuits*. This is the type most commonly found in the electrical installations that we have shown in the previous figures in this chapter. Other models, such as those indicated in Fig. 5, correspond to outlets with higher voltages and currents. The presence of the grounding conductor in all of them is notorious, which is an indication of the importance of electrical safety in the installation.

The NEMA specification is codified for all receptacles built in accordance with these standards, which include a wide variety of receptacle geometries.

Apart from those mentioned, we can mention the existence of specialized outlets, among which are the ground fault circuit interrupters (GFCI, for the acronym in English of Ground Fault Current Interrupter) and the outlets that have insulated gtound terminals. The former are used to prevent electric shocks and protect against electrocution, while the latter are suitable for the protection of electronic equipment sensitive to transient voltage spikes; that is, they are used to reduce electrical noise in these equipments. Both are easily distinguishable, as can be

---

* Typically a 15A plug will fit into a 20A outlet; however, a 20-amp plug will not, by configuration, fit into a 15-amp outlet.

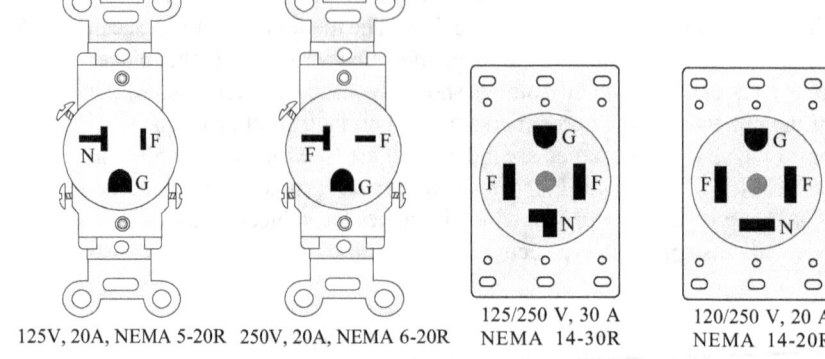

Phase (not used)

Phase

| 125V, 20A, NEMA 5-20R | 250V, 20A, NEMA 6-20R | 125/250 V, 30 A NEMA 14-30R | 120/250 V, 20 A NEMA 14-20R |

Fig. 5 Examples of outlets built to NEMA standards. The presence of the grounding terminal, denoted by the letter **G**, is notable. The phase and neutral terminals are denoted by the letters **F** and **N**, respectively.

seen in **Fig. 6**. The isolated ground outlet is generally orange in color and, according to electrical standards, it must be identified by a triangle, also orange, on its front face.

Of course, to take the energy that will feed the electrical devices, it is necessary to have the appropriate plugs that are inserted into the outlets. **Fig. 7** presents two of them. It is clear that there will be as many types of plugs as there are types of outlets.

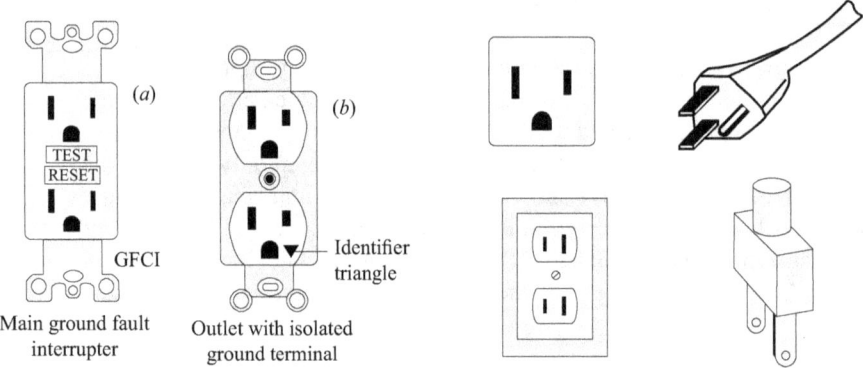

Fig. 6 (*a*) Ground fault interrupter. (*b*) Receptacle with isolated ground terminal. The triangle that identifies the outlet isolated from ground is observed.

Fig. 7 Plugs are the bridge between the outlet and the electrical equipment to be used. The shape of the plugs will depend on the type of outlet.

2. ELECTRICAL OUTLETS AND SAFETY

Did you know that outlets and plugs are the cause of a large number of fires and accidents, including deaths by electrocution in the use of electricity? In a typical electrical diagram, as presented in **Fig. 8**, all outlets derive their power from the main or service panel found in each residence, business or industry, and which, in turn, is energized by a reducing transformer, located outside the building.

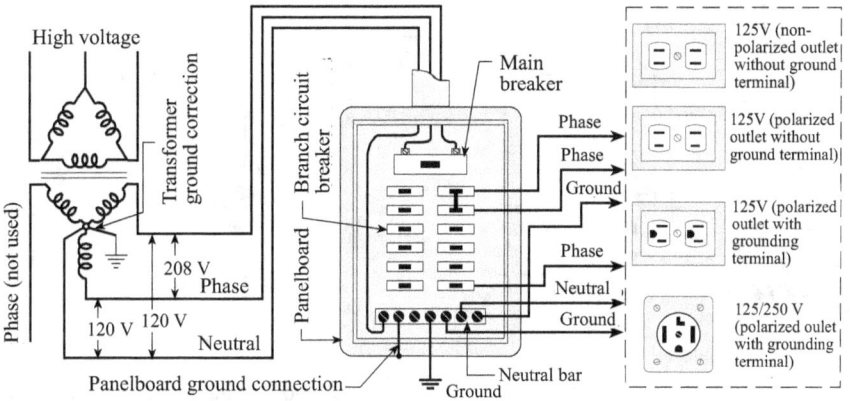

Fig. 8 Powering various types of outlets from the main board
of a residence.

Let's see how each type of outlet and its connection to the main panel can lead to a hazardous installation or a safe installation.

• *Non-polarized outlet without ground terminal*: Let's look at **Fig. 9**. The non-polarized outlet is connected between phase and neutral conductors coming from the main board. Its hot and neutral terminals are also connected to its hot and neutral slots. As it is a non-polarized outlet, with two slots of the same size, the plug, whose pins are also the same, can be connected in one position or another indistinctly. In this case the plug is inserted into the outlet in such a way that the phase of the power supply is connected to the phase conductor of the toaster which leads to its switch. When the toaster is turned off because its switch is in the open position, there is little possibility of an electrical accident inside the toaster, since the phase conductor is deactivated. The one-line diagram is shown on the right side of the figure. Suppose the plug is reversed in the outlet which is expressed in **Fig. 10**. The situation is now different since the neutral conductor is connected to the switch whereby the toaster is connected to the phase conductor (current conductor). phase on the one-line diagram). As a result, even when

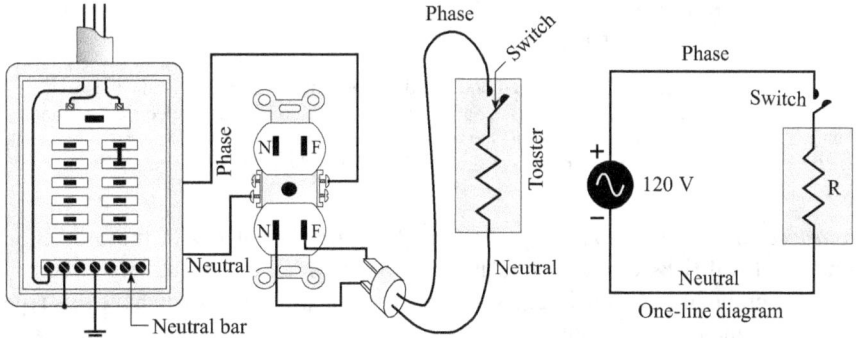

Fig. 9 Connection of a toaster to a non-polarized outlet and its one-line diagram. When the toaster is disconnected by opening the switch, the phase conductor is de-energized inside the appliance.

the switch has been turned off, there is a latent electrical hazard in the appliance.

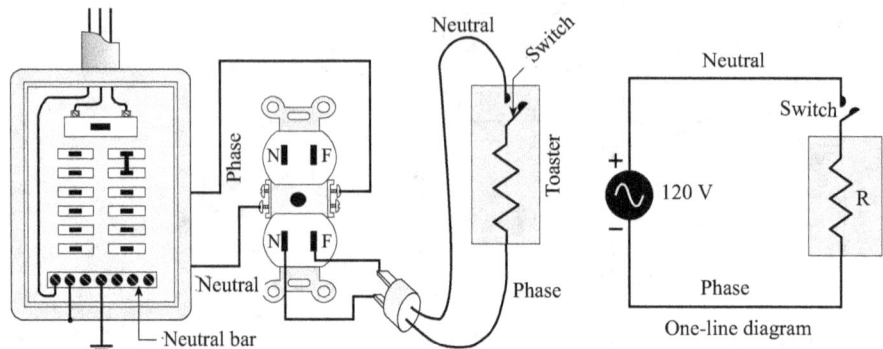

Fig. 10 Connecting a toaster to a non-polarized outlet and its single-line diagram. When disconnecting the toaster, opening the switch, the phase conductor remains active inside the electrical device.

• *Polarized outlet without grounding terminal*: This configuration is safer than the connected case because the presence of different size slots and pins in the outlet and plug makes it difficult to connect the equipment in a position other than that shown. That is the phase conductor will always correspond to the phase terminal of the switch and the neutral of the outlet will always be connected to the other terminal of the appliance. In this way the installation is less risky. The one-line diagram, shown in **Fig. 11**, will be unique.

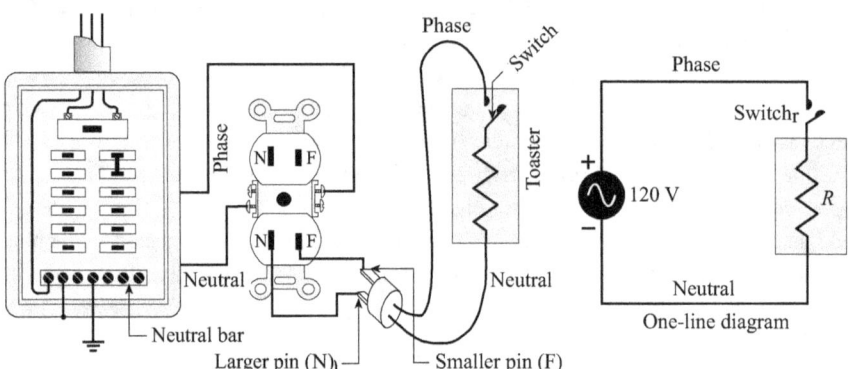

Fig. 11 Connecting a toaster to a polarized outlet and its one-line diagram. When the toaster is disconnected by opening the switch, the phase conductor is deactivated inside the electrical appliance. The only connection possibility corresponds to what is determined by the socket-plug combination, where the small slot (phase) is connected to the small plug and the large slot (neutral) to the large plug.

• *Polarized outlet with grounding terminal*: In order to understand the protection against electrical risks offered by an installation in which the outlets are polarized and have a grounding terminal, let's see what would happen when you have this type of outlet and the plug does not have the pin that connects an electrical appliance to earth, such as the one in **Fig. 12**. As suggested there, if the phase conductor comes into contact with the metal cover of the appliance, it will remain energized and the risk of electrocution could be high since the current would follow the path indicated by the little black arrows shown. A person who touches the metal cover will receive an electric shock that could be fatal. Current would flow from the board to the outlet and from there to the toaster, then through the person and finally back to the panel-board using the ground as a conductor. The one-line diagram is that of **Fig. 13** where the path followed by the current goes from the positive terminal of the source to its negative terminal.

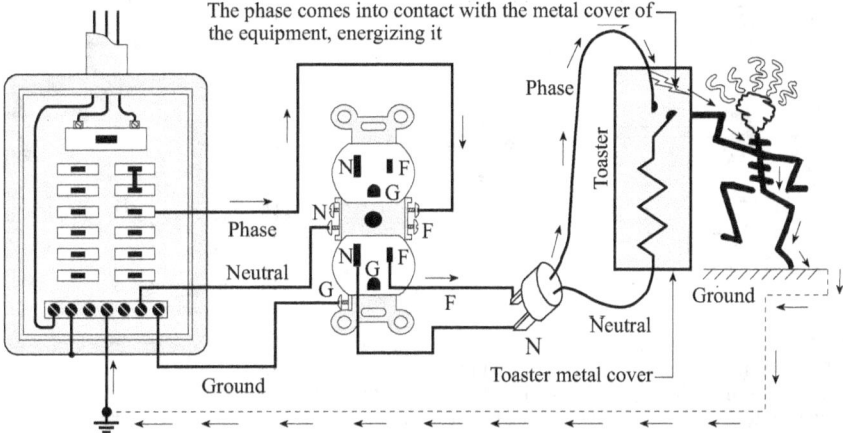

Fig. 12 Current path in a circuit when a contact fault occurs that connects the phase conductor with the metal cover of the appliance. The small black arrows indicate the path of the current that passes through the person.

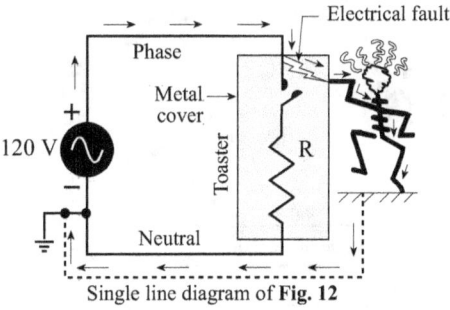

Fig. 13 Current path for **Fig 12** in case of contact of the phase with the metal cover of the electrical appliance when the outlet does not have the ground contact pin.

Let us now see what the situation is when the outlet is polarized and has a grounding terminal which is connected to the grounding terminal of the plug and to the metal co-ver of the equipment. We will refer to **Fig. 14**. If there is a contact between the phase conductor and the metal cover, the current has two paths to follow: through the person or through the metal and, from there, to the input board. The easiest way is through the metal part of the toaster since it has a much lower resistance than the human body. As a consequence, the current is returned to its source, the board, following the grounding conductor which is in contact with the metal part of the appliance. In this way the electric shock to the person is avoided and the risk of electrocution is eliminated. The excess current causes the breaker on the board to trip, disconnecting the phase conductor from the circuit. The small arrows indicate the flow of the current. **Fig. 15** shows the one-line diagram.

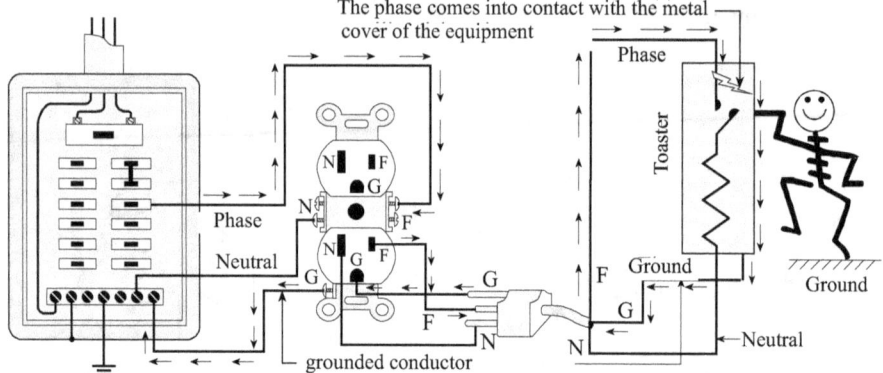

Fig. 14 Current flow in a circuit when a fault occurs that connects the phase conductor to the metal enclosure of the appliance. The current follows the path indicated by the small black arrows, which does not pass through the person. In this case the current is returned to the source through the grounding conductor causing the main panel breaker to trip. In this way the person is not in danger when the failure occurs.

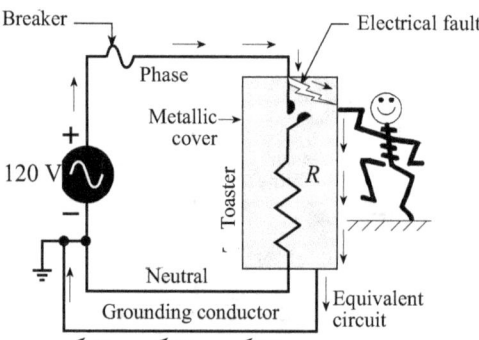

Fig. 15 Single-line diagram of **Fig. 14**. Current path, in case of contact of the phase to the metal cover of the electrical appliance, when the outlet and the plug have ground terminals and the metal cover is also grounded. Earth.

• *Polarized outlet with grounding terminal when the equipment connected to the branch circuit is 208 V:* In this situation (see **Fig. 16**) the outlet and the plug have two phase terminals, one neutral and one grounding. Also both the board and the washer have their metal casings grounded. If a circuit failure occurs, so that any of the phases comes into contact with the metal cover of the washing machine, the current will circulate through the grounding cable, avoiding the risk for people who come into contact with the cover.

of the washing machine. The panel breaker will trip in the presence of excessive current caused by the fault. The case is similar to that discussed for **Fig. 14**, except for the presence of two phases, either of which could cause the problem.

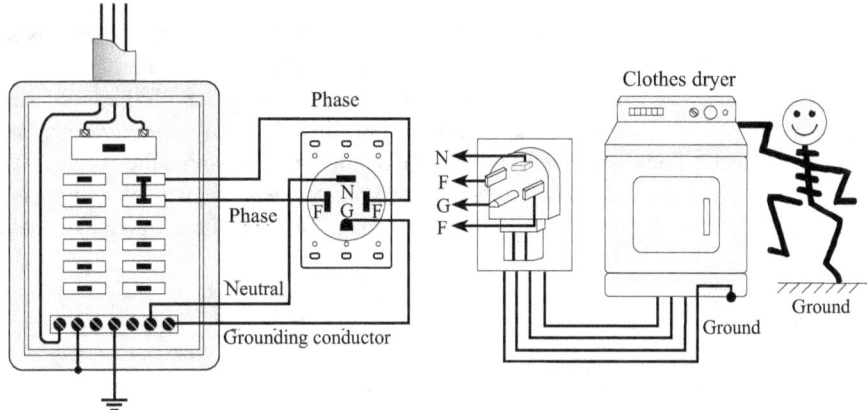

Fig. 16 Branch circuit for a dryer that is powered by two phases, a neutral, and the protective ground wire. If any of the phases came into contact with the metal frame of the dryer, a high current would be produced which would circulate through the grounding cable and would activate the corresponding switch on the main board. This would prevent a person touching the dryer from being electrocuted.

3. GROUND FAULT INTERRUPTER (GFCI)

Ground faults occur when the insulation of the active conductor (phase) breaks as a result of wear and tear due to excessive use, poor installation practices or improper use of equipment and appliances. If the insulation rupture occurs and the phase conductor makes contact with the metallic armor of the connected equipment a highly dangerous situation is created.

• *What is a ground fault*: The ground fault interrupter, commonly known as a GFCI (*Ground Fault Current Interrupter*), is an electrical device designed to detect ground faults. This interrupting capability protects people from fatal electrical shocks and prevents damage to buildings and property.

Normally the current flow in a circuit goes from the panelboard to the phase conductor; from there to the load and finally back to the panel through the neutral conductor. If this path is not followed and any of the phases comes in contact with the metallic cover of the load or with any other grounded surface or body such as a metallic conduit, water, concrete in contact with the ground or a lamp with conductive parts, we speak of a ground fault. A fatal situation could occur when a person is in this abnormal path. Hundreds of people perish annually in the world by electrocution. It is estimated that two thirds of these fatalities would not have died if the branch circuits where the ground fault occurred had been protected by GFCIs.

• *Effects of electric shock*: Electric shock in living beings causes damage that may be irreversible or fatal. Internal bleeding and destruction of muscle and nerve tissue may occur. This does not take into account the concurrent damage caused by falls, burns or bone fractures.

When a person enters the return path of the current to the source his body acts as another element in the circuit. The current through it will depend on internal and external conditions. The value of the voltage, the contact resistance with the electrified object, the duration of the contact, the path followed by the current in the body and the humidity conditions are, among others, the factors that determine the value of the current. The effects of current on the human body have been extensively studied and vary depending on whether it is a woman or a man. **Table 1** is a summary of these effects.

Effect on the human body	Current in mA (60 Hz)	
	Men	Women
Imperceptible.	0.4	0.3
Tickling, perception threshold,	1.1	0.7
Electric shock, no pain, no muscle contraction.	1.8	1.2
Electric shock, pain, no muscle contraction.	9.0	6.0
Electric shock, pain, muscle contraction threshold.	16.0	10.5
Electric shock, severe pain. Muscle contraction with immobilization. Respiratory faliure.	23	15
Muscle fibrillation after three seconds. Surely fatal.	> 100	> 100

Table 1 Effects of current on the human body.

Table 1 shows that a current greater than 100 mA is generally fatal, especially if its duration exceeds 3 seconds. Although the resistance of the human body changes according to internal and external conditions, it has been determined that it can vary from a few hundred to thousands of ohms. Its value at a given moment can be the threshold between life and death. This variability makes it more difficult to establish electrical safety conditions and a voltage that, although it can produce a tingling sensation in one person, can be fatal in another, depending on the humidity conditions of the skin, the main point of contact between a electrical equipment or appliance and the ground. The interior of the body is made up of water with mineral salts and other elements that are good conductors of electricity, so it is the skin that most affects the resistance values. The body's electrical resistance also varies according to the form of contact between the energized object and the skin: from hand to foot, between the two hands, between the two feet, etc.

When the skin is dry, its resistance can be as high as 500,000 Ω, its value being around 100 Ω when the skin is moist or soaked with water. The greatest injuries and electrocution occur in the latter case. The most dramatic case occurs in bathtubs where the fall of an electrical device such as a hair dryer could turn them into death traps. If we use the mentioned resistance values and a voltage of 120 V the current values will be:

$$I_1 = \frac{120}{500\ 000} = 0.24 \text{ mA} \qquad\qquad I_2 = \frac{120}{100} = 1\ 200 \text{ mA}$$

In the first case the current will be imperceptible while in the second it will prove fatal.

Generally, voltages above 50 volts are considered dangerous. From this value, measures must be taken to reduce electrical risks. It should be noted, however, that it is the current that produces important injuries and that its intensity in the human body must be limited to values that do not produce harmful effects.

• How does a ground fault interrupter work?: Once the reader has become aware of the importance of avoiding electrical shocks, let's see how a ground fault detector (GFCI) increases the safety of electrical systems. A GFCI is a ground fault interrupter that constantly monitors the difference in current between phase and neutral of a branch circuit. If this difference is not equal to zero, a breaker inside the GFCI trips and the branch circuit is disconnected from the network.

The operating principle is very simple and can be understood by looking at **Fig. 17** in which, to illustrate the operation, a simple resistor is used that represents an electrical device connected to the circuit.

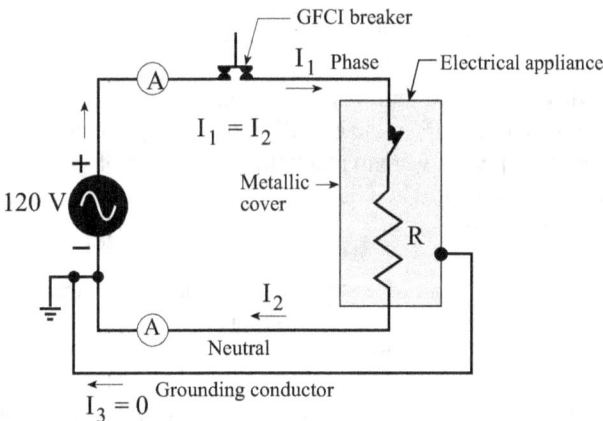

Fig. 17 In the circuit shown, the currents I_1 in the phase and I_2 in the neutral are equal. The current I_3, in the grounding conductor, is zero.

Under normal operating conditions, the ammeters, which measure the currents I_1 and I_2, have the same readings. That is, $I_1 = I_2$. The current in the grounding conductor, I_3, is equal to zero since there is no connection to the live (phase) conductor of the circuit. The switch external to the appliance which maintains the flow of current is closed guaranteeing normal operation.

What happens when a ground fault occurs inside the connected electrical appliance? This situation is illustrated in **Fig. 18**. The fault may be of such a nature that it does not necessarily cause the breaker to trip on the panelboard because the current is relatively small or because, if it does occur, its action time could be relatively large. This condition is represented by the resistance in **Fig. 18**, which connects the live conductor (phase) to the metal cover of the electrical appliance. As a result, the phase and neutral currents, I_1 and I_2, are not equal and a current originates in the grounding conductor. That is, between phase and neutral there is a current imbalance product of the ground fault. It is precisely this imbalance that the GFCI uses to activate the phase switch, as illustrated in that figure.

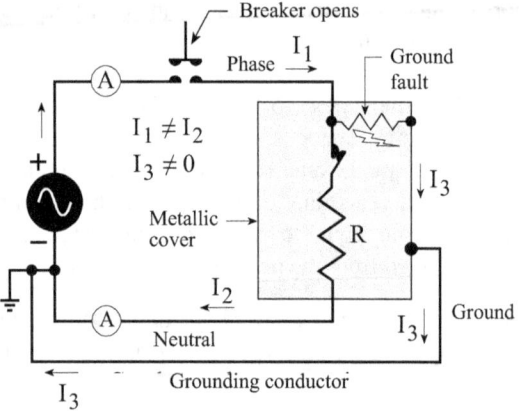

Fig. 18 When an earth fault occurs the currents I_1 in the phase and I_2 in the neutral are different. The current I_3 in the grounding conductor is not null.

Assuming that the ground fault was not a clear short circuit the current I_3 might not be strong enough to trip the main board breaker that protects the circuit. Under these circumstances if a person were to touch the metal cover of the device he would surely receive an electric shock. The GFCI would trip the breaker and trip the circuit, thus preventing a fatal outcome.

GFCIs are calibrated to trip when the current imbalance is greater than 5 mA and their response time is between 1/25 to 1/30 seconds, 25 to 30 times faster than the time between two successive heartbeats.

One might think that the protection offered by the panelboard switch and the presence of a grounding conductor would suffice to have a safe electrical installation. Again, it must be emphasized that these switches are designed for very high currents and that their fundamental role is to protect electrical installations, civil buildings and the equipment connected to them. GFCIs, on the other hand, are primarily intended for the protection of people who interface with those facilities. This protection, normally ignored in residential electrical installations, is the cause of unfortunate electrical accidents. Hence the need to create awareness in those who are responsible for the design and electrical wiring in homes.

To clarify how the GFCI protects a person from an electric shock let's look at **Fig. 19**. Let's assume that the current in and out of the GFCI is 1.25 A in the absence of any ground fault. If for any cause the active conductor (phase) comes in contact with the motor housing (metal enclosure) a fault occurs in the installation. If this contact does not create the current conditions to trip the breaker on the main panel, there is a latent danger to anyone touching the motor casing. For example, if the breaker is 30 A, it will not trip with a current difference of 0.25 A (250 mA). When a person touches the metal enclosure it adds an additional current path, which we will assume to be equal to 250 mA capable of producing serious consequences including death (see **Table 1**), if the duration of that current is sufficiently prolonged. This is the great advantage of the GFCI: when it detects that at the input terminal the current differs from the output current by 250 mA, it immediately disconnects the phase and neutral. Of course, this person will receive an instantaneous electric shock that may be reduced, due to its short duration, to a simple shock, but he will not be electrocuted.

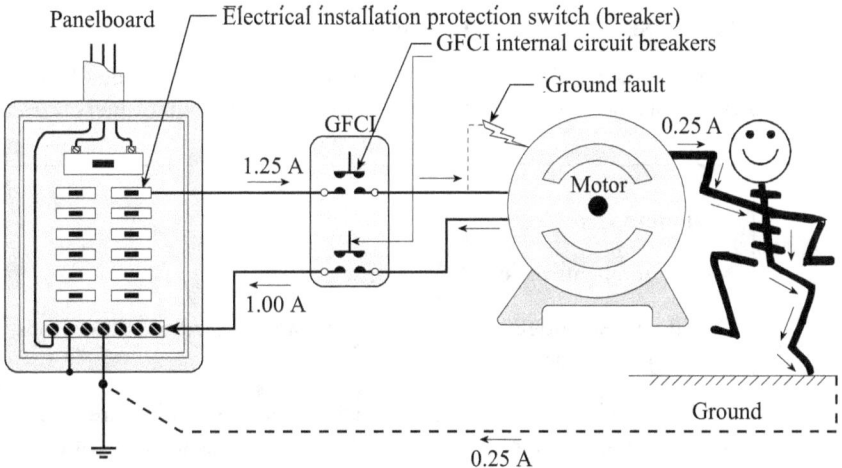

Fig. 19 The GFCI detects the difference between phase and neutral. When this difference reaches 0.25 A, for the figure shown, the device disconnects the phase and neutral conductors. In this way, the person receives only an instant shock when touching the motor housing.

4. CIRCUITO BÁSICO DE UN GFCI

The basic circuit of a GFCI is shown in Fig. 20. Its electronics are made up of operational amplifiers that detect the difference in the two coils of the differential transformer. This differential current is directed to an electronic comparator, which activates the trigger circuit when its output current is different from zero. In this way, both the phase and the neutral of the installation that supplies the load are disconnected. In order for the comparator output not to be equal to zero, it is necessary that a current is generated in the grounding conductor through the metallic cover of the connected load.

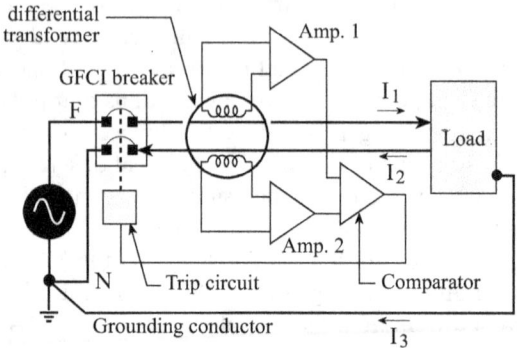

Fig. 20 Basic circuit of a GFCI. When a ground fault occurs the currents I_1 and I_2 are different, the current I_3 in the ground conductor is not zero and the GFCI trips.

5. WHERE SHOULD GFCIS BE USED?

To protect people, all 15A and 20A single-phase branch circuits located in the following areas must be connected to ground-fault circuit interrupters (GFCIs):

- Bathrooms.

- Environments outside the residences.

- All kitchen outlets subject to moisture and located on cabinet tops, islands and peninsulas.

- Sinks and places where laundry takes place and outlets are located above the top of furniture and within 1.83 m of the outside edge of the sink.

- House boats.

- Basements not intended for rooms and limited to storage or work areas.

- Garages.

- Areas near swimming pools and bathtubs.

GFCIs placed in external areas must be protected against environmental factors, specifically against water, so they must use plastic covers that prevent penetration into their interior.

A good rule of thumb for deciding on the use of GFCIs in residences is to know the status of the site where the facility's outlet is located. In general, if you are working in an environment where the presence of water is conspicuous and likely to create a path of little resistance for electrical current, this type of outlet should be used.

On the other hand, some appliances or receptacle outlet points do not require the presence of ground-fault current interrupters. We can cite the following exceptions:

- Inaccessible outlets, such as the one under the kitchen sink that feeds a garbage disposal.

- Outlet outlets placed on the inside of a garage ceiling and used to automatically open your door.

- Simple outlets in the kitchen, dedicated solely to powering a specific electrical appliance, such as a refrigerator or refrigerator.

Also, since GFCIs are very sensitive to the difference in currents at their input and output terminals, it is not advisable to place them in outlets that feed medical equipment, refrigerators, freezers, etc., whose operation does not allow an intermittent interruption of the power. electric service.

6. LIMITATIONS ON GFCI USE

It is worth warning about the limitations of GFCIs so as not to oversize expectations about them. These usage limits depend on your understanding of how it works. It should be emphasized that, under normal conditions, a GFCI trips only when a ground fault occurs, that is, when a fault return path to ground is created in the circuit. Some situations in which GFCIs do not provide personal protection or are unresponsive include the following:

- A GFCI does not protect against electrical shock when a person resting on a non-conductive surface touches two live (hot) conductors or one live (hot) conductor and the neutral. Under these conditions there is no return current through ground and the GFCI does not detect a difference in currents across its terminals, as indicated in **Fig. 21**.

- When a ground fault occurs involving a human being, they could receive a significant electrical shock. The advantage of a circuit connected to a GFCI is that this discharge, in its presence, occurs for a very short time, which prevents electrocution.

- A GFCI does not detect or trip when shorts occur between phase and neutral or between two phases, since the current is the same at its two terminals. In this case it is the branch circuit switch located on the main board, that must act.

- A GFCI does not trip when excess current occurs in the branch circuit. The protection must be provided by the corresponding switch (breaker) of the main board.

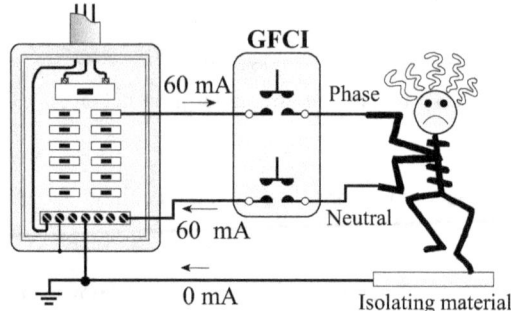

Fig. 21 If someone rests on an insulating material (dry wood, for example) and touches both the phase and the neutral at the same time, the currents in the terminals of the GFCI are the same and the circuit does not trip. As a consequence, the person receives an electric shock, 60 mA in this figure, which can be fatal.

Aside from the above limitations it should be noted that in some cases the GFCI may be subject to intermittent tripping, creating nuisance outages on the branches it protects. This is due to leakage currents along the circuit path which unbalances the GFCI. Based on this some manufacturers specify the maximum length of the branch circuit. On occasions these currents have their origin in the accumulation of moisture in the wiring, in outlets or lamp outlets. It is worth mentioning that the continuous tripping of the GFCI suggests the review of the electrical devices connected to the branchsince it could be defective equipment and, therefore, a latent electrical hazard.

7. GFCI TYPES

Outlets located outside of a residence, in bathrooms, in the garage, in work areas, and on top of kitchen cabinets must be GFCI protected. Also, 120V, 15 or 20 amp outlets within 6 feet of wash bowls must be GFCI type. There are three types of GFCIs available for residential use and described below.

- *Outlet Type*: This type of GFCI can be used to replace standard outlets often found in homes and apartments. It fits seamlessly into normal outlets and protects people from ground faults every time an electrical device or equipment is connected to the corresponding branch. When a ground fault occurs the receptacle-type GFCI disconnects both hot and neutral lines from the branch circuit. Its appearance is similar to that of normal outlets, except that it has buttons for test (test) and reset (reset) of the normal operating condition of the GFCI.

- *Panelboard type breaker*: This type of outlet is installed on the main panel of the electrical installation to protect specific branches. In this way, a ground fault that occurs in any of the outputs of the protected circuit will cause the interruption of the current. In this case, the GFCI disconnects only the phase of the powered circuit. As an added bonus, this type of GFCI trips when a short circuit or overload condition occurs. It has a single button, the test button. To restore the circuit (reset) the switch must be moved to the off position (off) and then to the on position (on). A GFCI of this class is more expensive than the outlet type.

- *Portable type GFCI*: When permanent GFCIs are not practical, portable GFCIs, which have a plug that plugs into a regular outlet, may be used. Some electrical appliances, like hair dryers, include GFCIs inside their cords.

All GFCIs should be tested periodically to verify that they are working properly to ensure the protection they offer when a ground fault occurs.

8. HEIGHT AND POSITION OF OUTLETS

The electrical codes do not establish strict regulations regarding the height of outlets above the finished floor. It is the conditions of use that determine the recommended height to place them. For example, a television placed at a certain height from the floor in a room indicates the possible location of the outlet that will power it. Likewise the location of outlets above heating units is not recommended, since any appliance connected by an electrical cord or extension cord could come into contact with the hot surface and cause a short circuit (see **Fig. 22**). **Table 2** presents typical outlet heights for different environments.

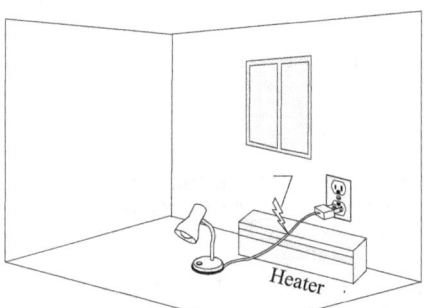

Environment	Height above the floor
General (rooms, corridors, etc)	30 cm
Kitchen gabinets	1 m to 1.15 m
Outdoors	45 cm
Garages	45 cm a 1.15 m

Tabla 2 Typical outlet heights.

Fig. 22 The outlet must not be placed above the heater because if the lamp cable comes into contact with its cover it could cause a short circuit as the insulation of the cable melts due to the heat..

The electrical codes do not specify the position of the ground terminal and, therefore, a socket can be oriented in any of the positions shown in **Fig. 23**. In the first case, the ground terminal is located on the lower part, while in the second it is located in the upper part.

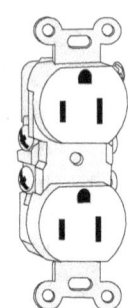

Fig. 23 The codes allow the orientation of a receptacle in any of the indicated positions.

Although there is no objection to the orientation of outlets in the standards, it is advisable to orient them in such a way as to minimize the chances of a short circuit when a three-prong plug is connected. The argument behind **Fig. 24**(a) is that if a metal cover comes loose and hits a plug that is not properly fitted to the socket, contact takes place between the metal of the plate and the ground, thus preventing a short circuit. In contrast, in **Fig. 24**(b), if a metal cap were to come loose, it would touch the live and neutral pins of the plug, resulting in a short circuit, spark, and possibly a fire.

A loose metal case could contact the ground terminal.

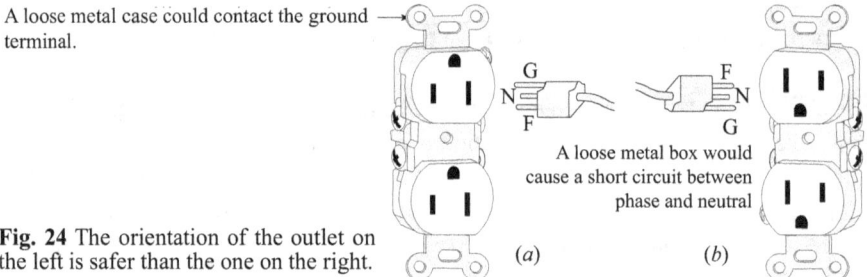

A loose metal box would cause a short circuit between phase and neutral

Fig. 24 The orientation of the outlet on the left is safer than the one on the right.

(a)   (b)

Fig. 25 presents two other possibilities for orienting a socket. As explained therein, the position of part (*a*) is safer than that of part (*b*).

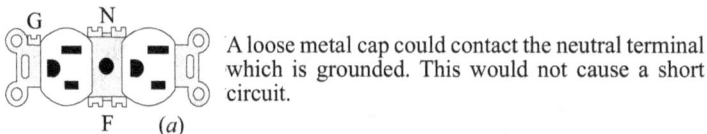

A loose metal cap could contact the neutral terminal which is grounded. This would not cause a short circuit.

For a multiconductor circuit the two parts of the outlet act independently when the metal bridge between them is broken. As a consequence, each section could be connected to 120 V, with a voltage of 240 V between the two parts.

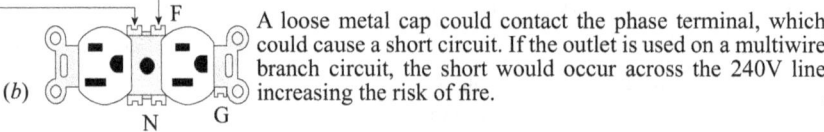

A loose metal cap could contact the phase terminal, which could cause a short circuit. If the outlet is used on a multiwire branch circuit, the short would occur across the 240V line increasing the risk of fire.

Fig. 25 The top outlet orientation is safer than the bottom outlet orientation.

9. MULTI-CONDUCTOR BRANCH CIRCUITS

A multiwire branch circuit is an electrical installation in which two different circuits use the same neutral. **Fig. 26** is a typical diagram of a multi-conductor circuit, connected to four outlets. It is observed that the two upper outlets are connected to one phase while the two lower ones are connected to the other phase. Neutral is common to all four outlets. That is, instead of using two conductors for the neutral, one for the upper outlets and one for the lower outlets, only one conductor is used, which leads to savings in the wiring of the installation. The grounding conductor is also common to both groups of outlets.

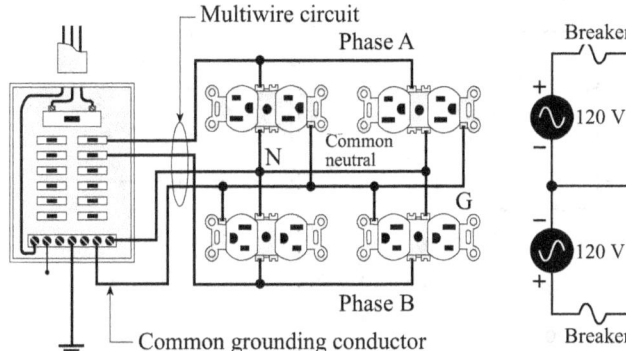

Fig. 26 Multi-conductor branch circuit   Fig. 27 One line diagram for **Fig. 26**.
(*multiwire*).

Suppose that in the multi-conductor circuit of **Fig. 26** there is a voltage between phase and neutral of 120 V and between phases of 240 V. The phase voltages are out of phase by 180°. If we keep only the two outlets on the left side and make a one-line diagram of **Fig. 26**, we arrive at **Fig. 27**. We will assume that those outlets supply the resistive loads R_1 and R_2. The ground wire has been omitted in order to simplify the explanation below.

The current in the neutral is the vector sum of the currents I_1 and I_2. If the loads R_1 and R_2 have the same magnitude, I_1 and I_2 are equal, but are out of phase by 180°. As a result, the current in the neutral will be equal to zero. This implies that there is no voltage drop in the neutral and, therefore, in a multiwire circuit the voltage drop is reduced with respect to a normal branch circuit. So not only is there a saving in conductors but energy loss is reduced. This reduction in voltage drop is also present if the loads represented by R_1 and R_2 are different. We must add that the reduction in the number of conductors translates into smaller size conduits.

Multiwire circuits are used not only for outlets, but can also power other types of loads, such as lighting fixtures. It should be noted that its use is not so common in homes.

Despite the advantages mentioned in relation to multiwire circuits, it is worth mentioning the dangers that underlie this type of installation. To get an idea of what we are saying, let's look at **Fig. 28**. An 800 W toaster is connected to one of the phases of the branch, while a 100 W electric lamp is connected to the other. The equivalent circuit is that of **Fig. 28(b)**.

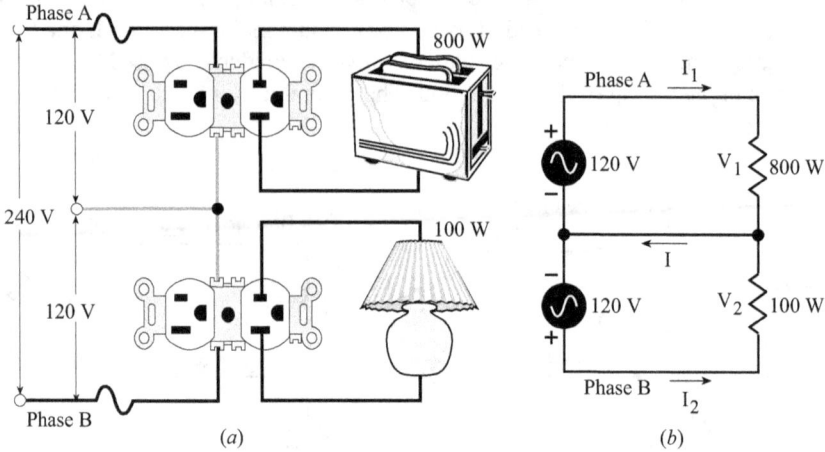

Fig. 28 Multiwire branch circuit to which loads of 800W and 100W are connected

The equivalent resistances of both devices can be calculated using the well-known relationship between power, voltage and resistance:

$$P = \frac{V^2}{R} \quad \Rightarrow \quad R = \frac{V^2}{P}$$

R values for the toaster and the lamp:

$$R_1 = \frac{120^2}{800} = 18 \qquad R_2 = \frac{120^2}{100} = 144$$

To highlight the weakness of the multiwire circuit, let us suppose that, for some reason, the neutral is disconnected, as we can see in Fig. 29. As a result, the current in the neutral is equal to zero and the voltage between the terminals of the two appliances is 240 V. The total current in the circuit is obtained by dividing the 240 V voltage by the sum of the resistances:

$$I = \frac{240}{18 + 144} = 1.48 \ A$$

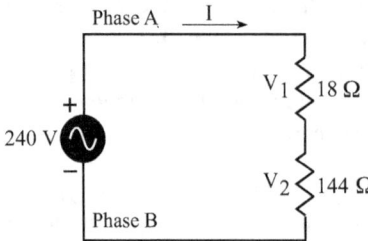

The voltages in the toaster and the lamp are given by:

$$V_{Tostadora} = 18 \cdot 1.48 = 26.64 \ V$$

$$V_{Lámpara} = 144 \cdot 1.48 = 213.12 \ V$$

Fig. 29 Single-line diagram of **Fig. 28(b)** when disconnecting the neutral.

From the calculations above we see that the voltage across the lamp exceeds its normal working voltage (120V). Consequently, it will be damaged. We can conclude that an open neutral in a multiwire circuit can cause irreparable damage to devices connected to the branch.

Another situation that could constitute a latent danger is the connection of a branch circuit, as indicated in Fig. 30, mistakenly thinking that in this way there is a multiconductor circuit.

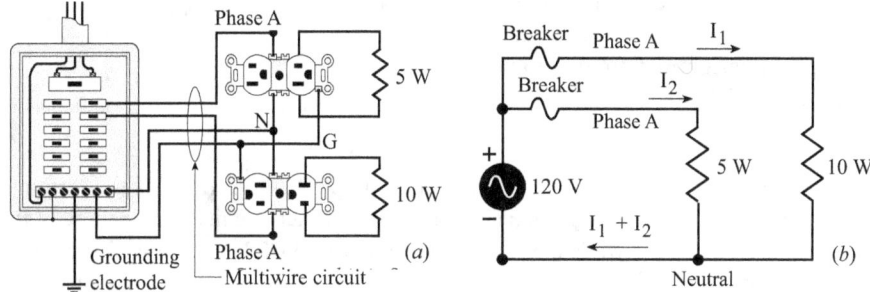

Fig. 30 When the load resistors are connected to the same phase, thinking that in this way a multiwire circuit is obtained, the neutral current can exceed the capacity of the conductor.

In this case, the two conductors that supply the 5 Ω and 10 Ω loads are connected to the same phase and, therefore, their voltage with respect to neutral is 120 V. The currents in the loads are:

$$I_5 = \frac{120}{5} = 24 \text{ A} \qquad I_{10} = \frac{120}{10} = 12 \text{ A}$$

The current in the neutral is the sum of the currents in the load resistances:

$$I_N = 14 + 12 = 36A$$

This current could cause overheating of the neutral and deterioration of its insulation, as well as potentially causing a fire.

To summarize, the advantages of a multiwire circuit are as follows:

 • There is a saving in the number of conductors used in the installations.

 • The voltage drop in the conductors is reduced.

 • Power loss in wiring is reduced.

Also, the disadvantages are these:

 • Possible damage to appliances due to disconnection of the neutral.

 • 240V voltage is present at the outlet boxes.

 • If a mistake is made in the connection, using a single phase instead of two different phases, the neutral can be overloaded.

A duplex outlet can be used on multi-conductor circuits. In this type of device, the two screws of the phase and neutral terminals are connected by a metal bridge that can be easily removed by cutting the small plate that joins them, as shown in **Fig. 31**. When the connection is broken between the phase terminal nuts, the top and bottom of the receptacle can be wired independently. The plate that joins the neutral screws is left intact. In this way two pieces of equipment that absorb large amounts of current can be connected to a multi-conductor circuit. It is common to refer to this connection as a split phase outlet.

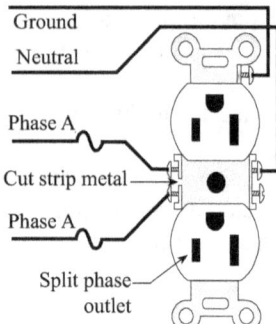

Fig. 31 A split phase outlet can be used on a multiconductor circuit.

In multi-conductor circuits, such as the one in **Fig. 31**, the screws of the terminal corresponding to the neutral cannot be used to connect other circuits to the neutral of the electrical installation. So, while the connection shown in **Fig. 32**(*a*) is allowed, the one in **Fig. 32**(*b*) is not allowed. The same applies to other electrical devices such as lamps. This measure makes it more difficult for the neutral to be loose in a multiwire circuit, since the connection to an appropriate connector inside the outlet box is more secure than the connection in the latter. The movements to which the outlets may be subject make them prone to the neutral becoming unstuck from the screw and causing it to disconnect with the consequences that we have already examined.

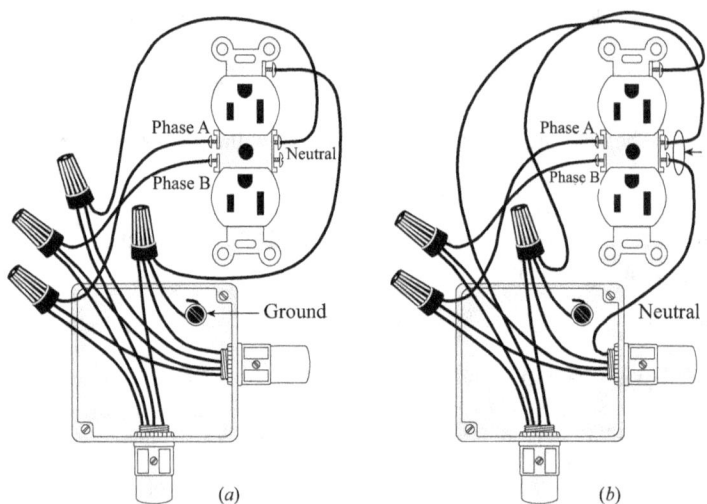

Fig. 32 The electrical codess do not allow making a connection to the neutral screws of an outlet when it comes to multi-conductor circuits.

Where a multiconductor circuit supplies more than one device or equipment on the same yoke or base, a means shall be provided to simultaneously disconnect the two phases at the location where the branch circuit originates. As indicated in **Fig. 33**, the protection in the power panel of the multiconductor branch circuit must be either two single-pole breakers, joined by a handle, or a two-pole breaker.

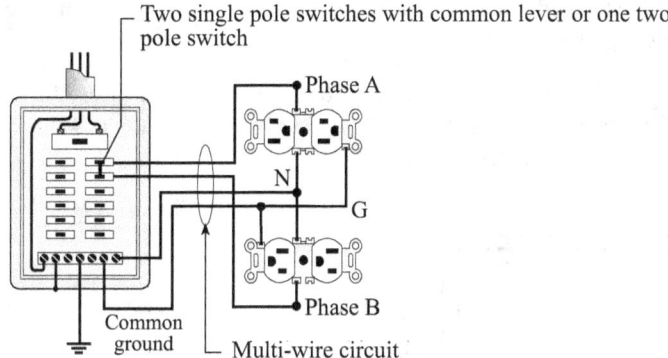

Two single pole switches with common lever or one two pole switch

Phase A

N

G

Phase B

Common ground

Multi-wire circuit

Fig. 33 In a multi-conductor circuit, a means of simultaneous disconnection must be used for both phases.

10. WIRING OF OUTLETS

In the wiring of 120 V outlets, several alternatives can be presented depending on the type of electrical installation.

Wiring an outlet: This is the most elementary case and corresponds to that of **Fig. 34**.

Wiring of two or more outlets: **Fig. 35** indicates what the electrical diagram would be like for two outlets. The scheme is repetitive for combinations of more than two outlets.

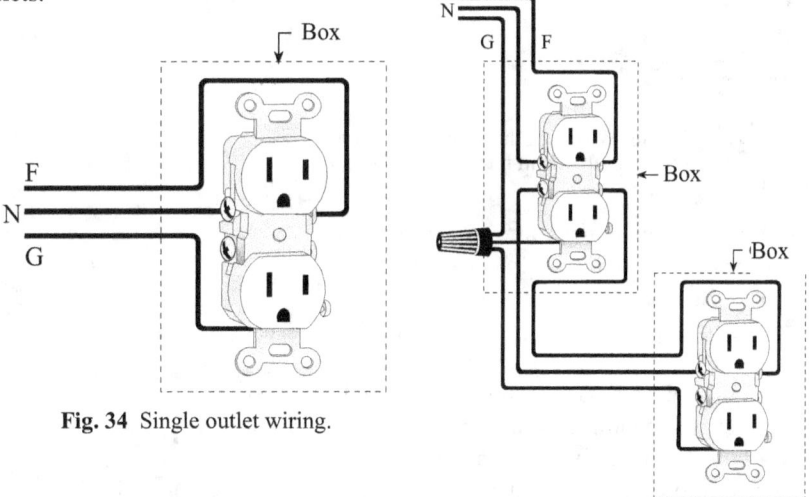

Box

F

N

G

Fig. 34 Single outlet wiring.

N

G   F

← Box

Box

Fig. 35 Two oulets wiring.

Wiring Three Outlets with Different Circuits: As shown in **Fig. 36**, two phases, one neutral, and the grounding conductor are used to power three outlets. The neutral is common to all of them. The phases alternate between the three outlets. The neutral and grounding conductor are common to all outlets. Phase F_1 feeds outlets 1 and 3, while phase F_2 feeds outlet 2. The scheme presented in **Fig. 36** can be easily extended to a larger number of outlets.

Wiring an outlet for 240 V appliances: This type of outlet consists of two phase conductors and a grounding conductor, as shown in **Fig. 37**. Note the ground connection of the box that houses the appliance. outlet.

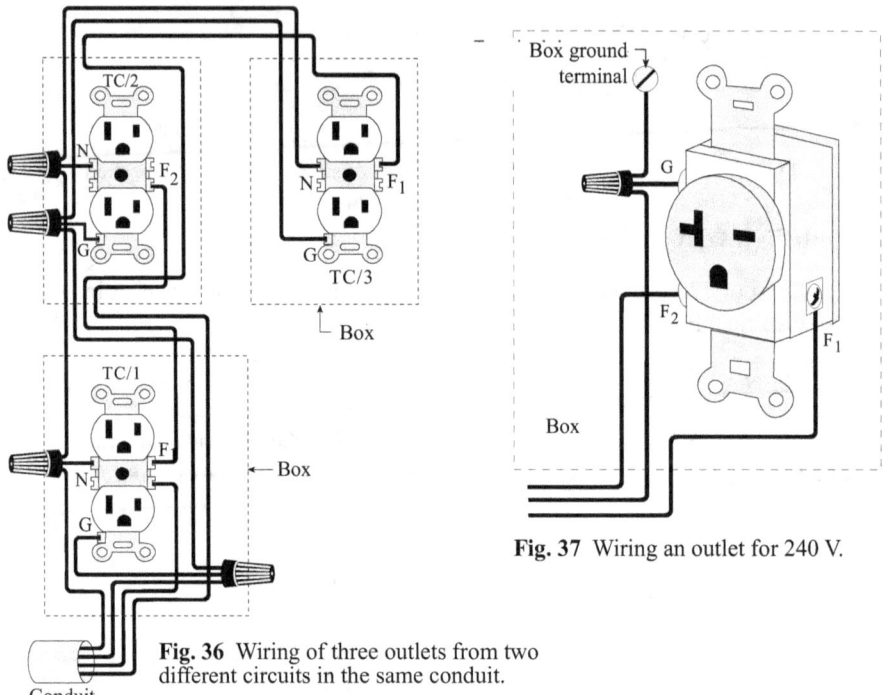

Fig. 37 Wiring an outlet for 240 V.

Fig. 36 Wiring of three outlets from two different circuits in the same conduit.

GFCI Wiring: Ground fault interrupters can be wired to protect individual appliances which is their most widespread use or to protect multiple pieces of equipment or appliances cascaded to the GFCI. **Fig. 38** corresponds to a schematic drawing of a GFCI. The device has the slots and the hole for the phase, the neutral and the earth, just as if it were a normal outlet. Two buttons, *test and reset*, allow you to check the proper functioning of the GFCI and restart its normal operation. Five screws are used to wire this device. The two upper ones, marked with the word *Line*, are used for the connection to the phase and neutral of the incoming power line of the GFCIs, as if it were a normal outlet. Used in this manner, each GFCI protects users from a single electrical appliance or

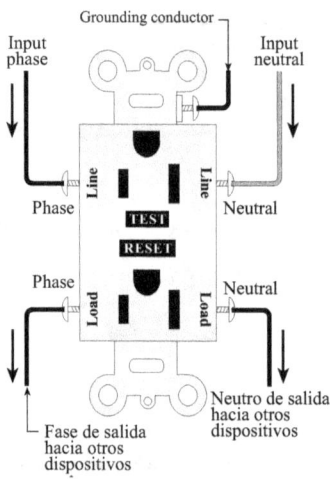

Fig. 38 Schematic drawing of a GFCI.

piece of equipment. The bottom screws, marked load, are used to protect other devices connected in cascade to a GFCI from ground faults. That is, these terminals lead to lamps, normal outlets, equipment, etc.

Wiring a GFCI and a regular outlet: See **Fig. 39**. The GFCI only protects the equipment or appliances connected to it. The regular outlet is not protected by the GFCI.

Wiring a GFCI to protect other devices, equipment and lighting: **Fig. 40** illustrates the connection. There you can see that the first device, a GFCI, receives power from the line at its terminals (line). The two regular outlets connect to the load terminals, and if a ground fault occurs on either of them, the GFCI trips.

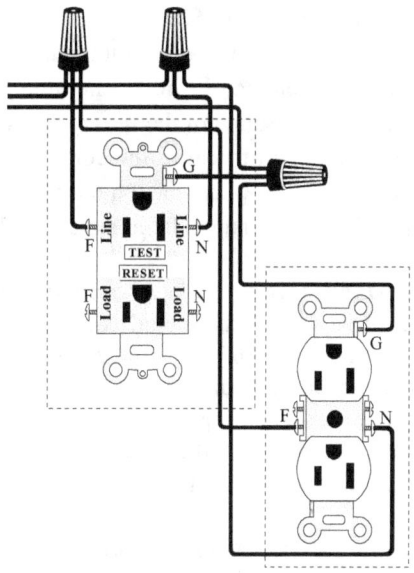

Fig. 39 Wiring a GFCI and a regular outlet.

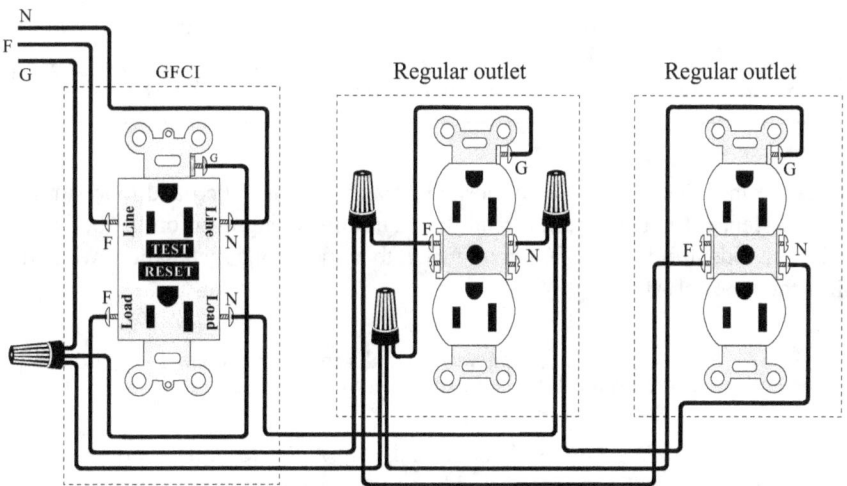

Fig. 40 Protection of two regular outlets by a GFCI. The latter detects any ground fault that may occur in the outlets and trips. Appliances connected to the GFCI are not limited to outlets, but could be light fixtures or any other device connected between the load terminals of the GFCI.

11. ARC-FAULT CIRCUIT INTERRUPTERS

When a phase makes firm contact with the neutral conductor or ground, the current generated is of such magnitude that it trips the protective breaker of the branch circuit. However when that contact is intermittent due to a poor connection or insulation failure of a conductor, a spark or arc is produced whose frequency could be such that the heat generated gives rise to very high temperpatures, in the range of thousands of degrees Celsius. The hot metallic particles expelled by the arc are enough to cause the combustion of many materials. *This is known as an arc fault.* If this phenomenon continues and there are combustible materials nearby (plastics, wood, paper, flammable liquids, etc.), unexpected fires could start.

Arc-Fault Circuit Interrupters: this type of interrupter disconnects the circuit if a situation occurs that could cause intermittent sparking and subsequent fires in a branch circuit. The AFCI differentiates sparks from normal circuit operations, such as connecting or disconnecting an appliance to the branch, from situations where intermittent sparking is caused by abnormal line behavior. The AFCI is designed to recognize the typical characteristic of a dangerous arc, detecting the rapid current fluctuations typical of this situation.

Due to the large number of fires caused by sparking some codes, including the **National Electric Code** (USA), establish the following:

> *All branch circuits supplying 125V single phase 15 and 20A loads in bed-rooms, dining rooms, living rooms, closets, hallways, or similar areas of dwelling units must be connected to arc fault interrupters to provide protection. to the complete branch circuit.*

AFCIs are installed the same way normal breakers are installed and look similar to a GFCI. Hence it is important to read the instructions, engraved on the body of the device, that identify it as a GFCI or an AFCI, to avoid wrong installation. We find the following types of AFCI:

AFCI for branch circuits and feeders: Installs in the main panel and protects the entire branch circuit. It is the most common type used today.

AFCI for electrical outlets: It is basically an AFCI to replace outlets and protects devices that are connected by plugs to said outlets.

AFCI-GFCI Combination: Use the advantages of an AFCI and a GFCI in a single device.

Portable AFCI: Similar to portable GFCI, it protects the cords that connect to the unit.

Some types of AFCI are designed to detect arcs which occur in both directions of the branch circuit toward the input and toward the output. Although AFCIs have a relatively high price it is expected that in the future awareness about the protection of human life will increase their use and lower their price.

12. SURGE PROTECTORS

Currently we find in use a wide variety of electronic equipment, such as computers, televisions, printers, microwave ovens, musical equipment, etc., which have electronic components sensitive to transient voltage spikes. These overvoltages can originate inside the facilities and are produced, among other equipment, by motors, copiers, laser printers, water heaters, electric stoves. They can also be caused by external factors, such as lightning strikes or rapid fluctuations in utility company voltages. In both cases voltage spikes can damage or cause equipment to malfunction.

Electrical codes pay attention to these voltage spikes and state that a *Transient Voltage Surge Suppressor* (TVSS) may be used for the protection of sensitive equipment. The connection diagram is illustrated in **Fig. 41**. Transient spike suppressors base their operation on the use of metal oxide varistors which absorb most of the energy present in the spikes. Only a part, with little destructive power, reaches the charge. The varistors act in less than 1 nanosecond, which guarantees safety against voltage spikes.

TVSS have the appearance of a single or multiple outlet and can be fixed or portable installed. Multiple TVSSs allow multiple computers to be connected at the same time. The number of devices to connect will depend on the capacity of the TVSS.

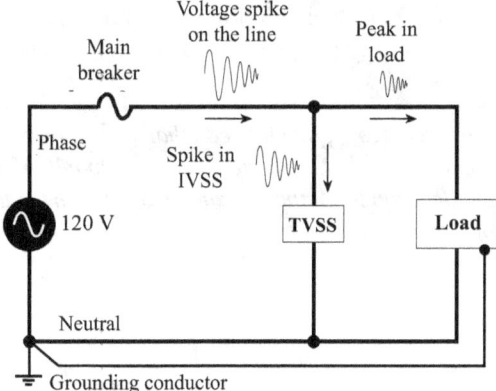

Fig. 41 Connection of a transient voltage spike suppressor to protect the load from its damaging effects.

13. CURRENT CAPACITY OF OUTLETS

Let us remember here that the branch circuits are designated according to the protection devices to which they are connected in the power panel. So, for example, a 20 amp branch circuit will be protected by a 20 amp breaker and a 40 amp branch circuit will be protected by a 40 amp breaker. It was stated earlier that, except for individual branch circuits, branch circuits will be 15 , 20, 30, 40 and 50 A. Likewise, we can affirm the following in relation to the current capacity of the outlets:

> *Outlets, as output devices, shall have a current capacity not less than the load to be served.*

For example, if an outlet powers a 12 amp load its minimum capacity should be 12 amps. Of course, these minimum values are subject to the current specifications of commercially available outlets. In this case a 15 amp outlet should be selected.

> *A single outlet, installed on an individual branch circuit, shall have an ampere capacity not less than that of that circuit.*

Thus, if an individual piece of equipment that consumes 35 amps is connected, the outlet must have a capacity of at least 35 amps.

> *When connected to a branch circuit supplying two or more receptacles or outlets, one receptacle shall not supply a load connected to it, by means of a plug and cord, in excess of the maximum specified in* **Table 2**.

As can be seen in that table, the maximum current in the load is limited to 80% of the capacity of the outlet. It can also be seen in **Table 2** that an outlet with a capacity of 15 A can be connected to a branch circuit protected by a 20 A breaker, as long as the maximum load to be connected, through a plug and cord, does not exceed 12A

> *When they are connected to a branch circuit that feeds two or more outlets or outlets, the current capacity of the outlets will correspond to* **Table 3**. *In the case of loads greater than 50 A, the current capacity of the outlet will not be less than that of the branch circuit.*

Circuit classification according to protection (A)	Outlet capacity (A)	Maximun load (A)
15 o 20	15	12
20	20	16
30	30	24

Table 2 Maximum current in loads connected to outlets by plug and cord.

Circuit classification according to protection (A)	Outlet capacity (A)
15	No greater than 15
20	15 o 20
30	30
40	40 o 50
50	50

Table 3 Current capacity of outlets according to the type of circuit.

La **Fig. 42** recoge lo establecido en cuanto a normas sobre la capacidad que deben tener los tomacorrientes.

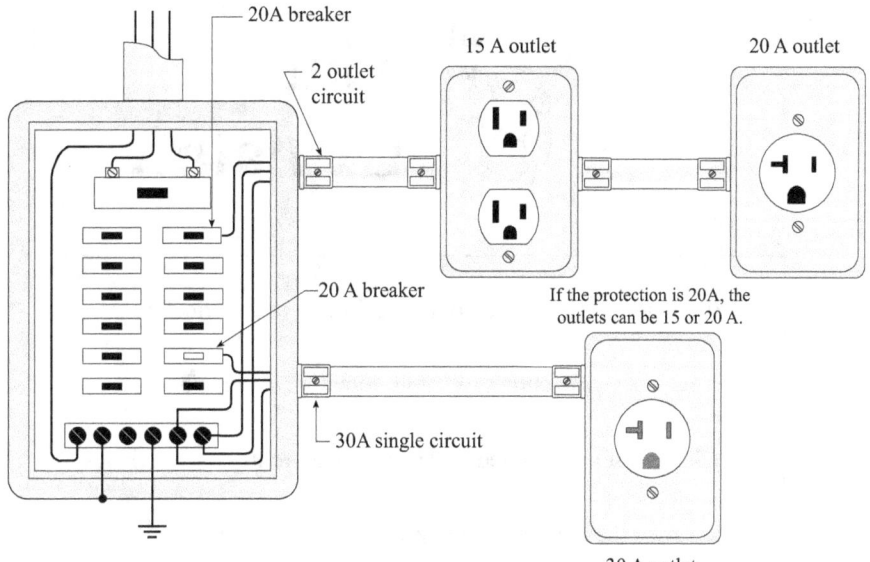

Fig. 42 Capacity that the receptacles must have in relation to the classification of the branch circuits.

14. SYMBOLS USED TO REPRESENT OUTLETS

When an electrical installation is designed, electrical plans are used to indicate where the different elements that make it up are placed. It is then necessary to have the symbols that represent the outlets, the lamps, the switches, the boards, etc. In the case of outlets, **Fig. 43** illustrates the most commonly used symbols. It should be emphasized that, in any case, the designer of the installation must include in the plans the list of symbols used to represent the elements of the installation. Anyone interpreting or studying the plans should refer to that list in order to make a safe installation.

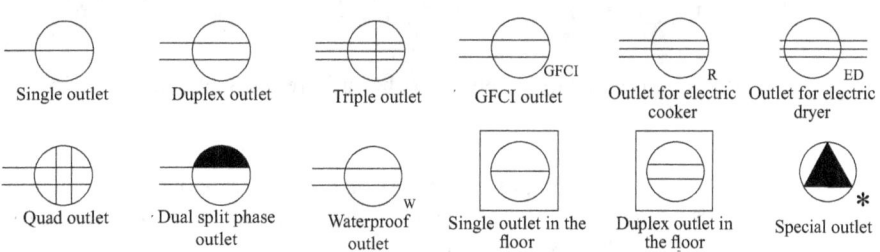

* **Special outlet**: the asterisk can be replaced by a letter such as a, b, c, or by a set of letters such as DW, ED, to indicate dishwasher and electric dryer. Other symbols may also be used, the meanings of which must be specified on the installation drawings.

Fig. 43 Symbols to represent outlets for different uses.

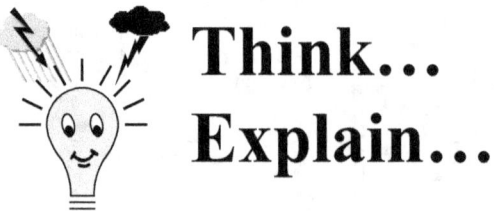

Think...
Explain...

---

1. State the definition of an outlet according to electrical codes.

2. What is a single outlet? What is a double outlet? What is an individual outlet?

3. What is a non-polarized outlet and what are its characteristics?

4. Describe the features of a duplex, polarized, grounding-type outlet.

5. In a polarized outlet, does the smaller slot correspond to the hot or neutral conductor?

6. How do you indicate on an outlet that it is a quality device?

7. Why must outlets on a branch circuit have a ground terminal?

8. From an electrical point of view, how are outlets classified?

9. How are ground fault circuit interrupters and ground isolated interrupters identified?

10. Explain in detail the electrical hazard present when non-polarized outlets are used.

11. Explain the operation of a polarized outlet when connected to a plug and how the geometry of both elements improves the safety of an electrical installation. Refer to **Figures 9** and **10**.

12. Explain why a polarized outlet with a grounding pin is unsafe when connected to an outlet that does not have a grounding pin. Please refer to **Fig. 12**.

13. Explain, referring to **Fig. 14**, why a polarized outlet with a grounding terminal offers safety to people in the event of a contact between the phase and the metal enclosure of the equipment.

14. What is a ground fault? When does a ground fault occur and how can it be fatal to a person?

15. What is a ground fault interrupter (GFCI)?

16. List the factors that determine the value of the current through a person when he or she comes into contact with an energized conductor.

17. What is the range of variation of the resistance of the human body?

18. Describe the effects of current magnitude on the human body.

19. Of the electrical effects on a person, which factor is more important: voltage or current?

20. Describe how a ground fault detector works.

21. How does the GFCI compare to a normal breaker in terms of speed of response and the current value that trips them?

22. Is it true that, even with the use of a GFCI, a person can receive a relatively high shock and not be electrocuted? explain.

23. Describe the operation of the basic circuit of a GFCI based on **Fig. 20**.

24. Under electrical codes where in a residence is the use of GFCIs required?

25. Name the abnormal conditions on a branch circuit where a GFCI is involved that do not cause the GFCI to trip.

26. What could be the causes of intermittent operation on a GFCI?

27. Describe the different types of GFCIs studied in this book.

28. At what current values does a GFCI typically trip?

29. If a person comes into contact with the hot and neutral of a branch circuit that is protected by a GFCI, will the GFCI trip?

30. Can a defective outlet, which does not have a ground terminal, be replaced with a GFCI? Explain.

31. What restrictions are established regarding the height of the location of electrical outlets above the floor?

32. Mencione las alturas típicas de los tomacorrientes sobre el nivel del piso.

33. Although the orientation relative to the floor that an outlet must have is not regulated by the electrical codes, explain why an orientation where the outlet's ground hole is facing up is safer.

34. What is a multiwire circuit? Draw a diagram to help understand your explanation.

35. Explain the advantages and disadvantages of a multiwire circuit relative to two branch circuits serving different loads.

36. Describe how a receptacle can be used on a multi-conductor circuit (split-phase receptacle).

37. Why is it not allowed to make splices in the neutral screws of an outlet in multi-conductor circuits?

38. What should be the means of disconnection for a multiconductor circuit?

39. From the schematic diagram for a GFCI, explain the use of the different connection buttons and screws.

40. What is an arc fault interrupter (AFCI)?

41. How does an arc fault interrupter work?

42. Where should arc fault interrupters be used? Describe the different types of AFCIs.

43. What is a Transient Voltage Surge Suppressor (TVSS)?

44. How does a TVSS work?

45. According to what is established in the codes, what is the maximum load that can be connected to an outlet using a plug and cord?

46. Can a 15 amp outlet be part of a branch circuit protected by a 20 amp breaker?

Practicing your knowledge

1. Is there an electrical risk in **Fig. 44** for those who touch the metal cover? explain.

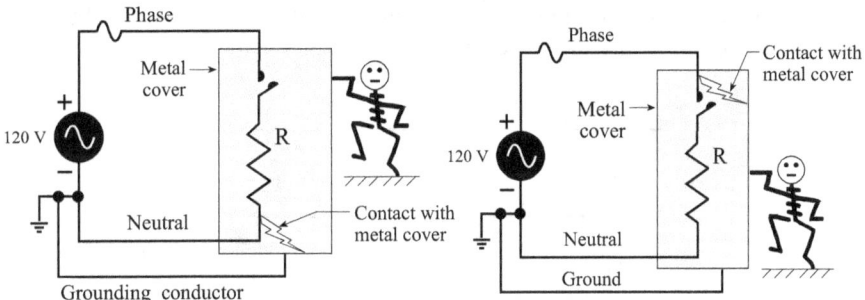

Fig. 44 Exercise 1. **Fig. 45** Exercise 2.

2. Is there an electrical risk in **Fig. 45** for those who touch the metal cover? Explain.

3. Is there an electrical risk in **Fig. 46** for those who touch the metal cover? Explain.

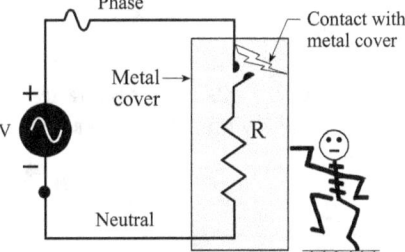

Fig. 46 Exercise 3.

4. A GFCI is designed to trip at a current of 5mA. What is the ground fault resistance that will draw that current for a voltage of 120V?

5. In **Fig. 47**, the grounding conductor has been mistakenly confused with the phase conductor. What is the electrical risk?

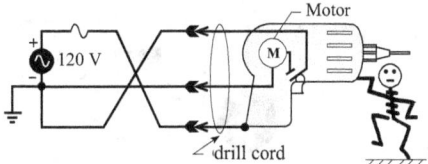

Fig. 47 Exercise 5.

6. In the multiconductor circuit of **Fig. 48**, electrical devices of 1,200 W and 600 W and 120 V are connected. *a*) Calculate the

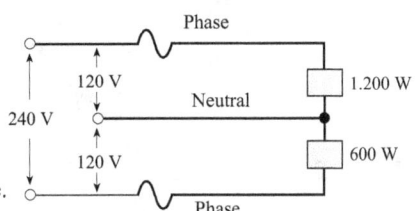

Fig. 48 Exercise.

current in each appliance. *b*) If the neutral were disconnected, what would be the voltage in each appliance? Would any of them be damaged?

7. Two 800-W toasters are connected to a 120-V multiwire circuit. If the neutral were accidentally disconnected, would either of the toasters be damaged?

8. In the multiconductor circuit of **Fig. 49**: *a*) Determine the current in each of the loads. *b*) Determine the current in the neutral. *c*) If the neutral is disconnected, what are the currents and voltages in the loads?

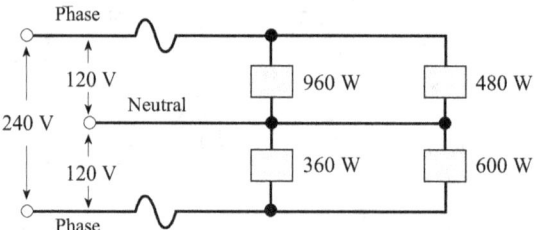

Fig. 49 Exercise 8.

9. As described in this chapter, a GFCI has a test button (TEST) that allows you to verify its proper operation. In **Fig. 50** the test circuit has been added. Explain how the latter works according to what you have studied.

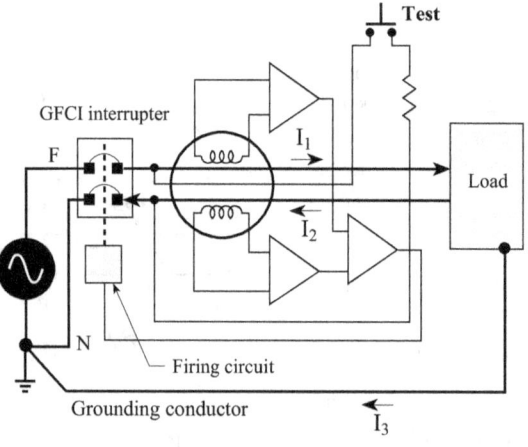

Fig. 50 Exercise 9.

INTERRUPTORES

15. GENERAL ASPECTS ABOUT THE SWITCHES

Electrical installations make intensive use of manual electromechanical switches, operated at will, when you want to connect or disconnect an element of the circuit, mainly lamps or lighting bulbs. **Fig. 51** corresponds to the image of a common electrical switch.

> *In the above context, a switch is a manually operated device to interrupt the current that feeds an electrical load.*

Switches are basically binary elements: they are either fully open or fully closed, there is no intermediate position. This characteristic: open or closed, has been extended to other types of switches used in a wide variety of applications. Thus switches have been designed that act by pressure, differences in level, temperature, flow, etc. Another category is made up of electronic switches which are widely used in electronic equipment.

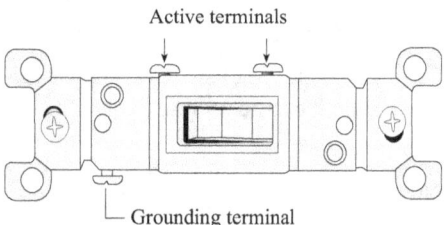

Fig. 51 Electromechanical switch used in electrical installations to turn a lamp on or off, or to activate a load from a single location.

Common switches have the basic structure of **Fig. 52**. The mechanism consists of a metal sheet that, by sticking or detached from the point of contact, connects or disconnects, respectively, the current that goes to the load. The manual movement of the switch lever between the connected or disconnected positions (ON and OFF) causes the displacement of the metal sheet, connecting or disconnecting the energy that feeds the load. Of course, the lever is made of an insulating material to prevent accidents from occurring due to electrical contact.

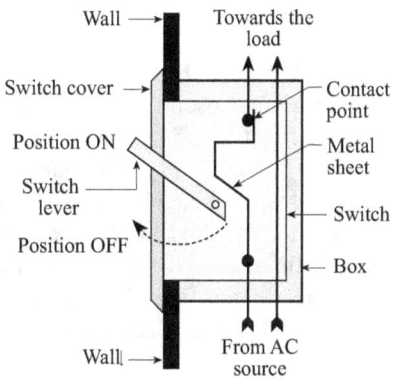

Fig. 52 Basic structure of a switch.

16. TYPES OF SWITCHES

The most basic type of switch is the so-called *blade switch*, currently in disuse and whose application is limited to demonstration purposes or high power industrial applications. Although its structure is very simple, as shown in **Fig. 53**(*a*), it can be used to illustrate the operation of any other type of switch. In that same figure, the circuit symbol is presented, which is used to represent the blade switch.

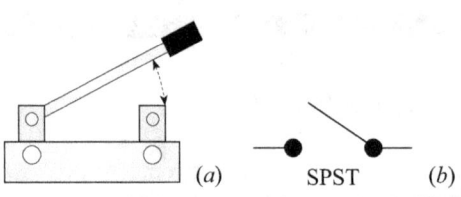

Fig. 53 (*a*) Basic structure of a SPST switch (one pole, single throw). (*b*) Schematic diagram.

Fig. 54 Elementary circuit to turn on a lamp by means of a SPST switch (single pole, single throw).

The blade switch in **Fig. 53** (*a*) is known as a single pole single throw switch and is commonly referred to as a SPST (Single Pole, Single Throw) switch. Its diagram is shown in **Fig. 53**(*b*). The current is cut off when the switch opens and feeds the load when the switch closes. An elementary circuit with this type of switch is shown in **Fig. 54**.

Another type of switch is known as SPDT or *single Pole, Double Throw switch*. A blade switch can be used to explain its operation. For this we will refer to **Fig. 55**.

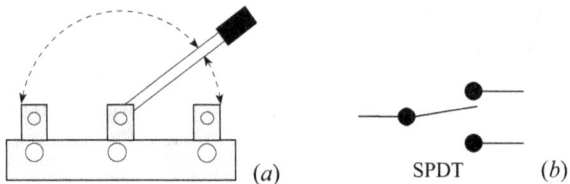

Fig. 55 (*a*) Basic structure of a SPDT switch (one pole, double throw). (*b*) Schematic diagram.

It is observed in **Fig. 55** (*a*) and (*b*) that the switch has two different positions and that the blade can oscillate around the midpoint interconnecting the central point with the ends. This type of switch is known as a *three-way switch*. A more correct designation would be that of a *three-terminal switch* since, really, there are two ways that are activated when the mobile arm moves from one point to another. However, custom has prevailed and those who work in electrical installations have imposed the the expression "three-way switch." **Fig. 56** illustrates the circuit of a SPDT switch to control two different lamps from the same position. Later we will study how three-way switches are used to control lamps from two different positions.

Lamps L_1 and L_2 light up when the SPDT switch is in positions 2 and 3, respectively.

SPDT switches have a common terminal for the two circuits (terminal 1 in **Fig. 56**) and two terminals that connect to the two lamps they control, which are known as traveling terminals (terminals 2 and 3 in **Fig. 56**). In a three-way switch there is no identification of the ON or OFF position, since its two positions can be used to turn a lamp on or off. In residential electrical installations the most widespread use of this type of switch is to turn a lamp on or off from two points far from each other. This situation occurs in cases such as the following:

1. When you have a lamp in the middle of a ladder and you want to turn it on when you start to climb the ladder and turn it off when you reach the top.

2. When you want to turn on the light in a room at the entrance and turn it off from the bed. Three-way switches are placed near the door and next to the bed.

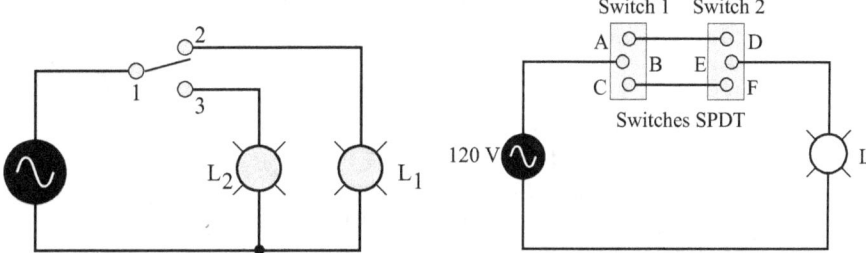

Fig. 56 Elemental circuit to turn on two lamps by means of a SPDT switch.

Fig. 57 Diagram for turning on a lamp using two SPDT switches.

Let's look at **Fig. 57**, where a lamp is controlled by switches 1 and 2. To understand how the circuit works suppose that switch 1 is located at the bottom of a ladder and switch 2 at its top. When points B and C of switch 1 are connected and switch 2 is in the upper position (D and E connected), lamp L is off (**Fig. 58**). If someone approaches the bottom of the ladder and passes switch 1 up, connecting points A and B (**Fig. 59**), a current is established in the circuit through the BADE path and the lamp lights up.

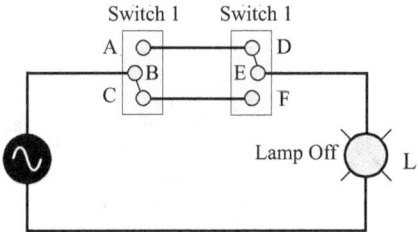

Fig. 58 Control of a lamp from two places: switch 1 is in the lower position, switch 2 is in the upper position and the lamp is off.

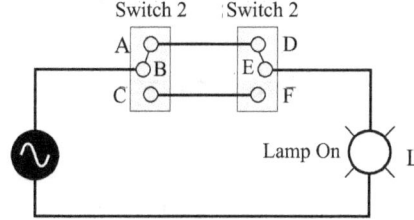

Fig. 59 Control of a lamp from two places: switch 1 is in the up position switch 2 is in the up position and the lamp turns on.

When the person reaches the top of the ladder, he turns off the lamp, turning switch 2 down, which leads to Fig. 60. As can be seen, the circuit is broken by connecting points E and F.

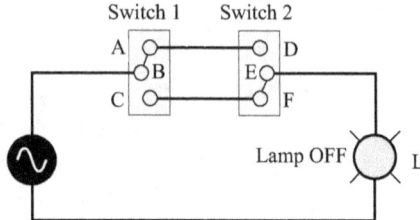

Fig. 60 Control of a lamp from two places: switch 1 is in the upper position, switch 2 is in the lower position, the circuit opens and the lamp turns off.

In the circuits of **figures 58** to **60** it can be seen that the mechanism is repetitive, and the lamp can be turned on and off from any end of the ladder.

In reality, a three-way switch is presented as shown in **Fig. 61**. The switch has four terminals: those used for wiring the installation (the common terminal and the traveling terminals) and the connection terminal. to ground The common terminal is distinguished from the traveling terminals by having a dark color.

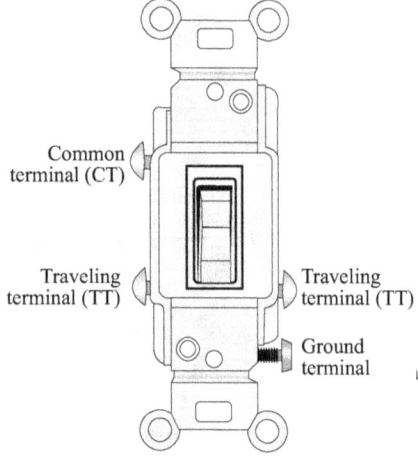

Fig. 61 Three way switch. The common terminal is usually dark, while traveling terminals are colored.

Finally, we must mention the four-way switches which used in conjunction with two three-way switches allow a lamp to be turned on or off from three different places. **Fig. 62** corresponds to the two positions that a four-way switch can adopt.

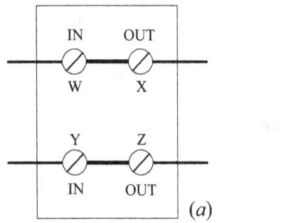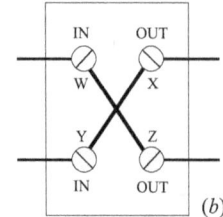

Fig. 62 Ways to connect a four-way switch: *a*) Input and output are directly connected (W to X and Y to Z). *b*) The input and output are cross-connected (W to Z and Y to X).

The two input terminals (IN) are on the left and the output terminals (OUT) are on the right. **Fig. 63** is a picture of a four-way switch. The two input terminals and the two output terminals have different colors. The ground terminal must be green.

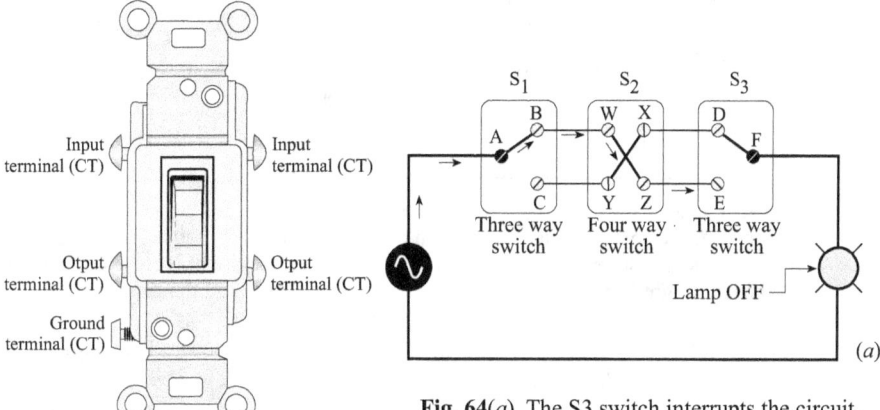

Fig. 64(*a*) The S3 switch interrupts the circuit.

Fig. 63 Four way switch. The input and output terminals are marked by different colors.

When it comes to controlling a lamp from three different points, two three-way switches and one four-way switch must be used. The latter must be placed between the three-way switches. For this purpose, we will refer to **Fig. 64** to illustrate how a four-way switch works in conjunction with two three-way switches. In **Fig. 64**(*a*) the lamp is off, since the switch S_3 interrupts the circuit.

The lamp can be turned on or off from any switch. Starting from the previous figure, **Fig. 64**(*b*) shows how the lamp is turned on from S_1. When the mobile arm of the switch S_1 passes from point B to point C the closed circuit indicated by the arrows is established. If later any of the moving arms of one of the switches changes position, the lamp will go out when the circuit is opened.

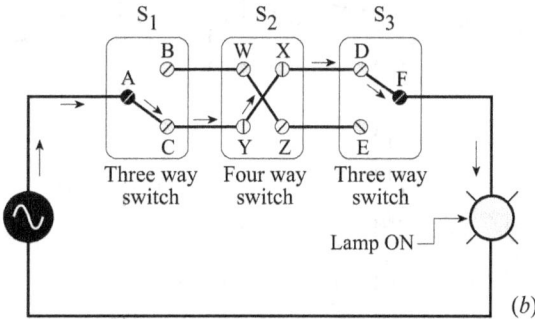

Fig. 64(*b*) When the moving arm of S_1 passes from point B to point C, the circuit is closed and the lamp lights up. Further displacement of the moving arm of any switch will open the circuit and the lamp will go out.

Starting again from **Fig. 64**(*a*), the lamp can be turned on from the four-way switch, S_2. This follows from **Fig. 64**(*c*), as we can see below in the following figure.

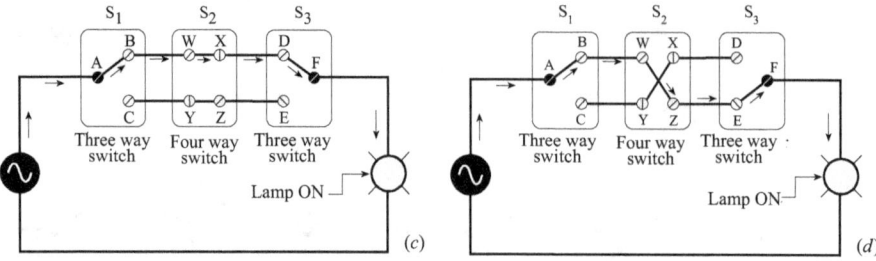

Fig. 64(*c*) When S$_2$ is actuated, points W and X are linked, as are points Y and Z. As a result the circuit is closed following the path ABWXDF, and the lamp lights up.

Fig. 64(*d*) We start from **Fig. 64**(*a*). When S$_3$ is actuated, points F and E are joined. As a result, the circuit is closed following the path ABWZEF, and the lamp lights up.

By activating the switch S$_2$, the points W and X are connected and the path ABWXDF is created, along which the current flows and the lamp lights up. Finally, from **Fig. 64**(*d*), it can be seen how the lamp is turned on by switch S$_3$. When the mobile arm passes from point D to point E the ABWZEF path is established, which allows the current to flow and the lamp to turn on.

17. SWITCH WIRING. PICTORIAL DIAGRAMS.

The conductors enter through a SPST switch: **Fig. 65** shows this case. Power enters the switch first and goes to the lamp it controls. Note the use of connectors to splice the conductors into the switch and light housings. The physical layout of the wiring is indicated in the same figure.

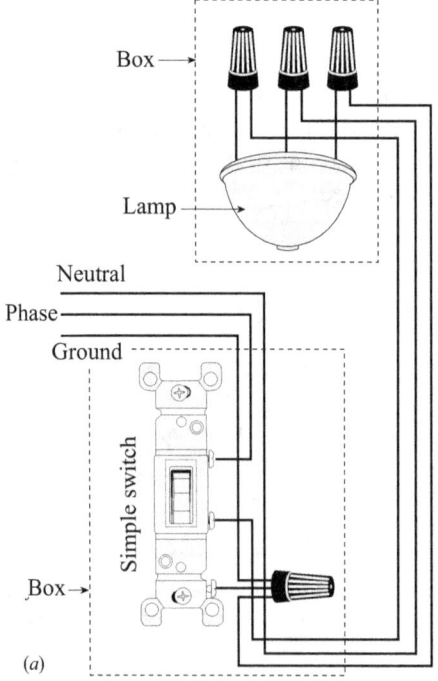

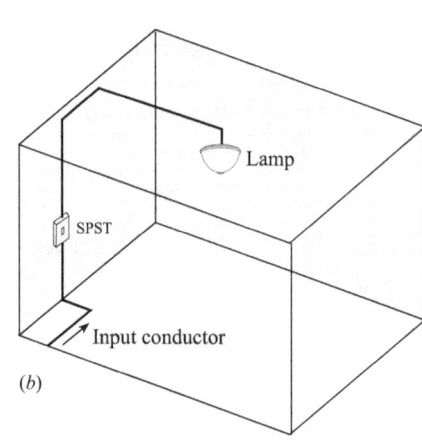

Fig. 65 (*a*) Wiring a single switch (SPST) when power leads enter the switch. (b) Pictorial diagram of the installation in a room.

The conductors enter through the lamp controlled by a SPST switch: In **Fig. 66** this case is indicated. Since the power enters through the lamp, the phase conductor must go directly to the switch and then return to the lamp.

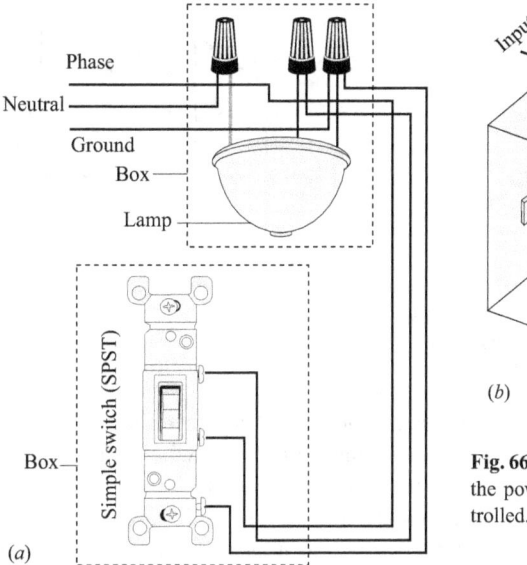

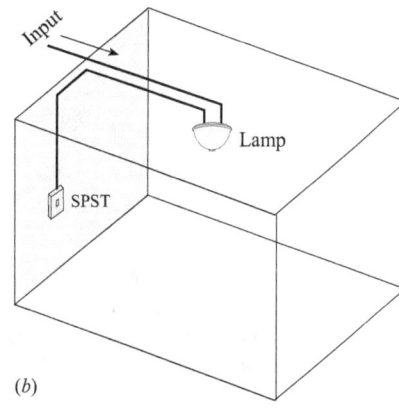

(b)

Fig. 66 (a) Wiring of a single switch (SPST) when the power conductors enter the lamp to be controlled. (b) Pictorial diagram of the installation.

(a)

Wiring two 3-way switches to control a lamp when power is coming through one of the switches and the lamp is at the end of travel: Mistakes are often made in the wiring of 3-way switches. Hence the importance of paying attention to the connection of the conductors to these switches. The traveling terminals are connected to each other as shown in **Fig. 67**. One of the common terminals is connected to the input phase, while the other common terminal goes to the lamp. A pictorial diagram of the installation is presented in **Fig. 68**.

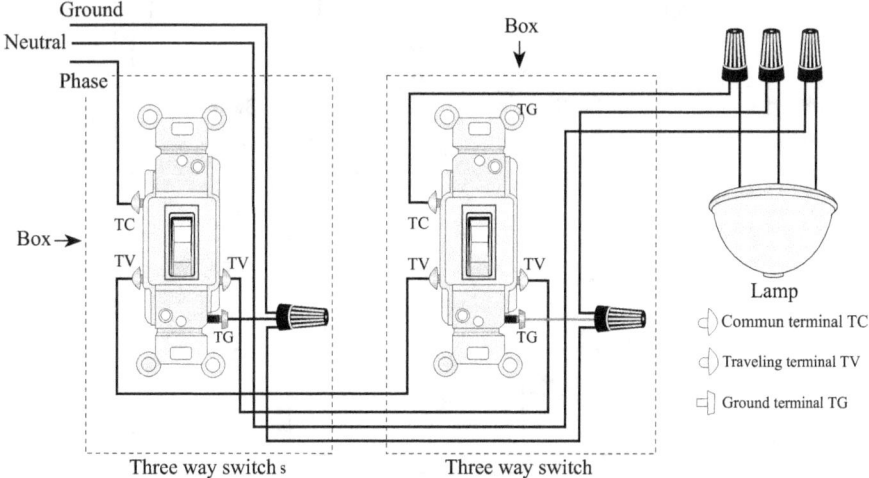

Fig. 67 Control of a lamp by means of two three-way switches.
Power enters through a switch.

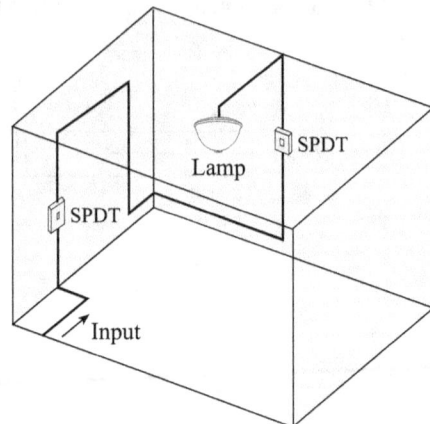

Fig. 68 Pictorial diagram of the installation for the control of a lamp from two different points, through the use of three-way switches. Power enters through one of the switches.

Wiring of two three-way switches to control a lamp when power enters the lamp: Wiring is shown in **Fig. 69**. The phase goes directly to switch 1 to connect to its common terminal. Also, the common terminal of switch 2 is connected directly to the lamp. The grounding conductor is distributed throughout the installation connecting to the lamp and all switches. The traveling terminals of the three-way switches are connected together as shown in **Fig. 69**. The neutral enters the lamp and does not extend to the switches. The pictorial diagram corresponds to **Fig. 70**.

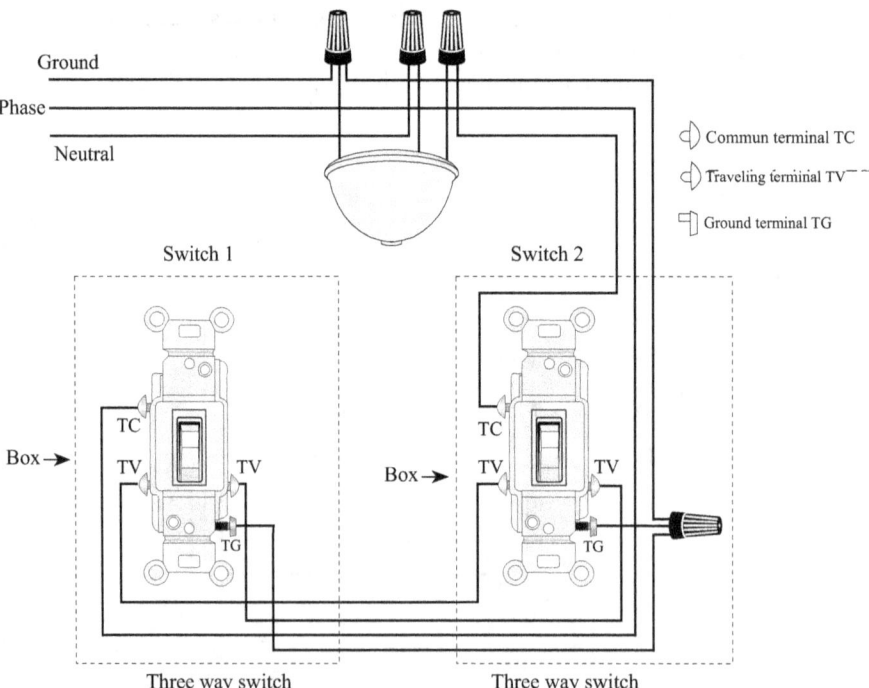

Fig. 69 Control de una lámpara mediante dos interruptores de tres vías.
La alimentación entra por la lámpara.

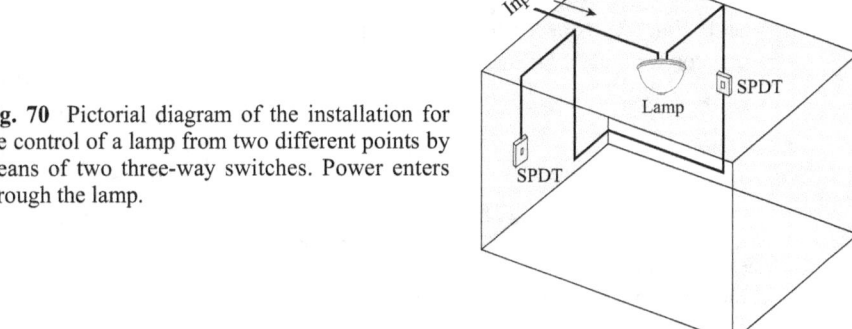

Fig. 70 Pictorial diagram of the installation for the control of a lamp from two different points by means of two three-way switches. Power enters through the lamp.

Wiring of two three-way switches to control a lamp when power enters the lamp: Wiring is shown in **Fig. 69**. The phase goes directly to switch 1 to connect to its common terminal. Also the common terminal of switch 2 is connected directly to the lamp. The grounding conductor is distributed throughout the installation connecting to the lamp and all switches. The traveling terminals of the three-way switches are connected together as shown in **Fig. 69**. The neutral enters the lamp and does not extend to the switches. The pictorial diagram corresponds to **Fig. 70**.

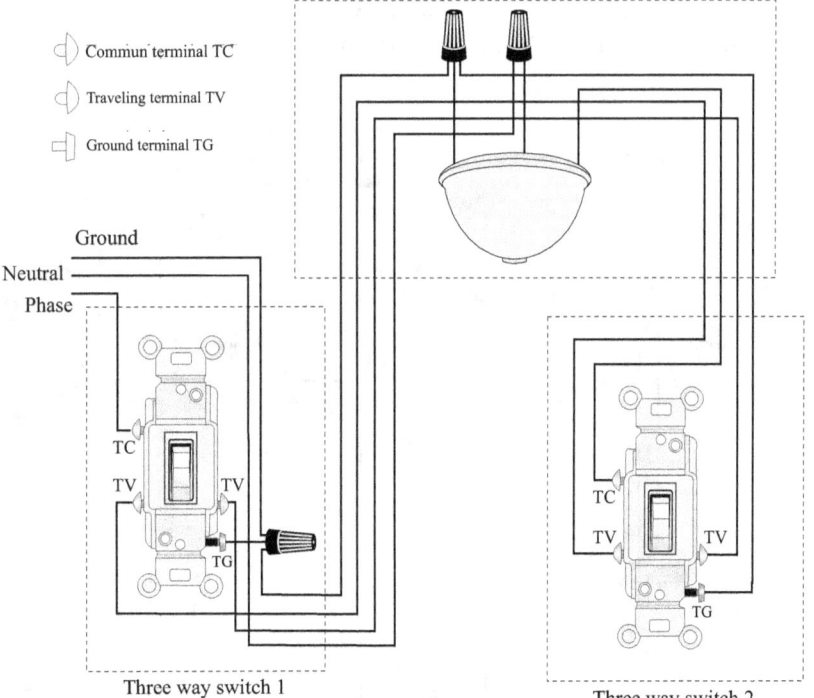

Fig. 71 Control of one lamp by two three-way switches. feeding go through a switch.

En la **Fig. 72** se observa el diagrama pictórico que corresponde a la **Fig. 71**.

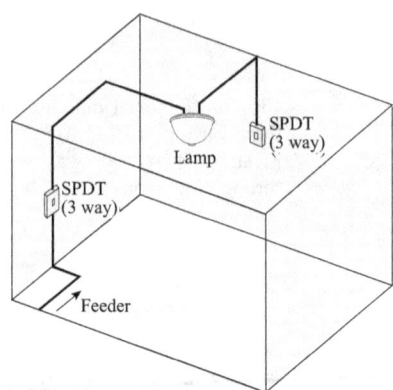

Fig. 72 Pictorial diagram of the installation for the control of a lamp from two different points by means of two three-way switches. Power enters through a switch and the lamp sits between the two switches.

Wiring two 3-way switches and one 4-way switch to control a lamp from three different locations: To turn a lamp on or off or to turn power on and off to an outlet from three different locations, one 4-way switch can be used, placed between two three-way switches, as shown in **Fig. 73**. As can be seen, the traveling terminals of the three-way switch S_1 are connected to the lower terminals of the four-way switch S_2. The upper terminals of S_2 connect to the traveling terminals of S_3. The phase connects to the common terminal of S_1 and the common terminal of S_3 goes to the lamp. The neutral conductor is not connected to any terminal of the switches: it goes directly to the lamp, after passing through their boxes, where connectors are used to make the splices.

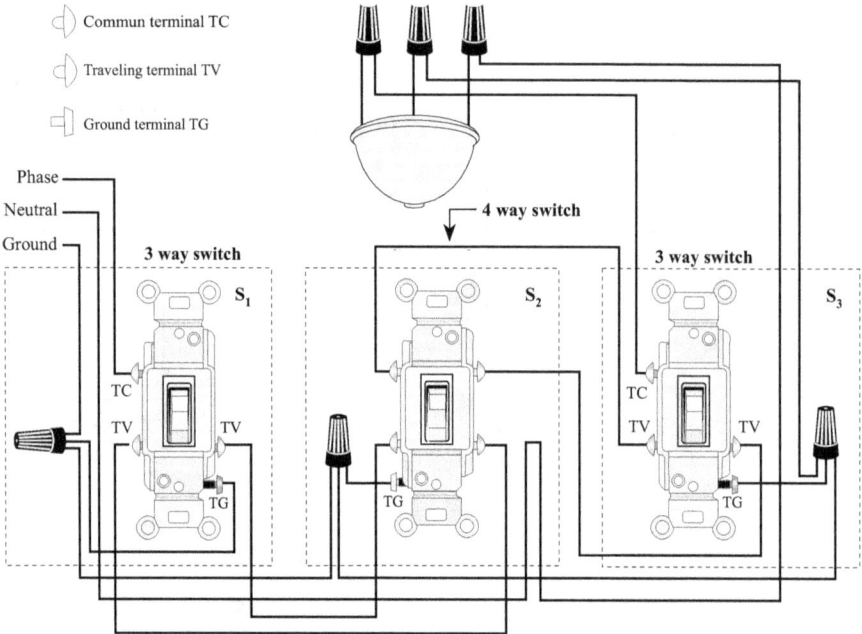

Fig. 73 Control of a lamp from three different places by means of two three-way switches and one four-way switch. Power enters through switch S_1.

When the lamp is controlled from more than three different points, more four-way switches are added between the three-way switches as the simplified diagram in **Fig. 74** shows.

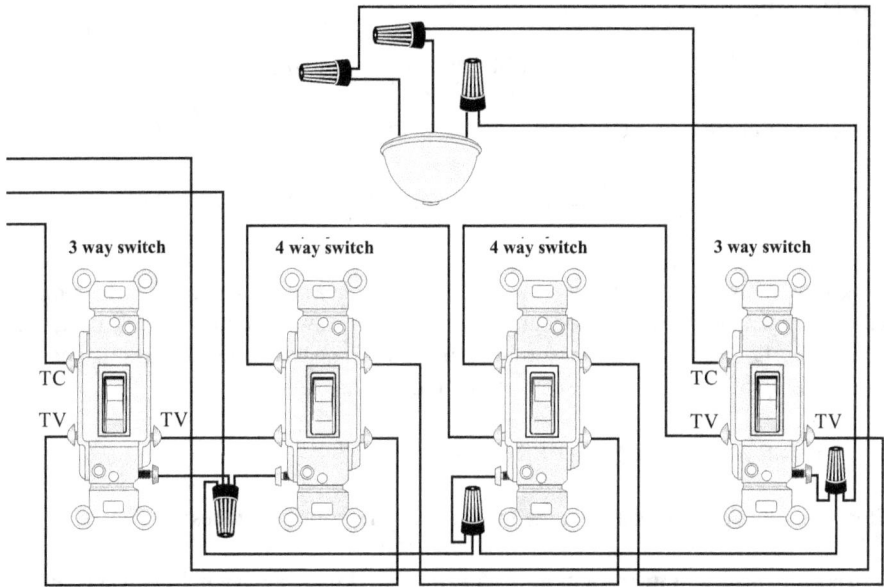

Fig. 74 Control of a lamp from four different places by means of two three-way switches and two four-way switches.

18. SPECIFICATIONS OF THE SWITCHES

The switches used in electrical installations are specified according to the voltage of use and the maximum current that can pass through them. For such purposes, the following can be mentioned:

Marking: Circuit breakers must be marked with the maximum current and voltage values and sometimes the hp that it can support.

Also, when using a switch, the following should be taken into account:

(1) *The resistive or inductive current, including that of discharge lamps, must not exceed the maximum duty cycle of the switch for the operating voltage.*

(2) *Tungsten filament lamps must not absorb a current greater than the maximum current supported by the switch at operating voltage.*

(3) *The current in switch-controlled motors must not exceed 80% of the maximum switch current at rated operating voltage.*

In relation to point (2) above, it should be noted that tungsten lamps draw a large current when turned on. This is because, initially, before heating up, the resistance of the tungsten filament is relatively small compared to its resistance when the lamp reaches its normal working temperature. Typical values of this resistance for a 100 W and 120 V light bulb are:

$$R_{Caliente} = 144 \ \Omega \qquad R_{Frío} = 10 \ \Omega$$

The current in the two previous cases is:

$$I_{Caliente} = \frac{120}{144} = 0.83 \ A \qquad I_{Frío} = \frac{120}{12} = 12 \ A$$

indicating a higher current when the bulb begins to heat up. Even though the temperature change from hot to cold occurs very quickly, the switch contacts must be able to handle the 12 amp current that is initially present. In general, the switches that certify the handling of this current are marked with the letter T.

19. OTHER SWITCH CONSIDERATIONS

Here are some points regarding the proper and safe use of switches:

Three-way and four-way switches: Three-way and four-way switches must be wired so that the entire connection and disconnection process is done on the active phase conductor. Where metal raceways or metal armored cables are used, wiring must be done to prevent induction heating of the metal.

According to the previous aspect, the conductors involved in the connection and disconnection of energy will be connected to the active conductor (phase).

Induction heating in ferrous metal ducts* occurs when the currents that circulate in the raceway in both directions (entering and leaving) are not compensated as their magnetic fields do not cancel. Two ways (correct or incorrect) of wiring three-way switches to

> It is important to gain experience with wiring 3-way and 4-way switches. It is common to find electrical installations where the connections between these switches, being poorly made, do not perform their function correctly.

---

* Heating also occurs in cables with metal armor.

control a lamp are shown in **Fig. 75**. Assuming that ferrous metal ducts are used, the one in **Fig. 75**(*a*) does not produce induction heating, since the current flowing in one direction is equal to that flowing in the opposite direction. However in **Fig. 75**(*b*) there will be heating in the ducts since the currents, when dealing with a single conductor, create uncompensated magnetic fields. In both figures the grounding cable and the connectors in the boxes have been omitted.

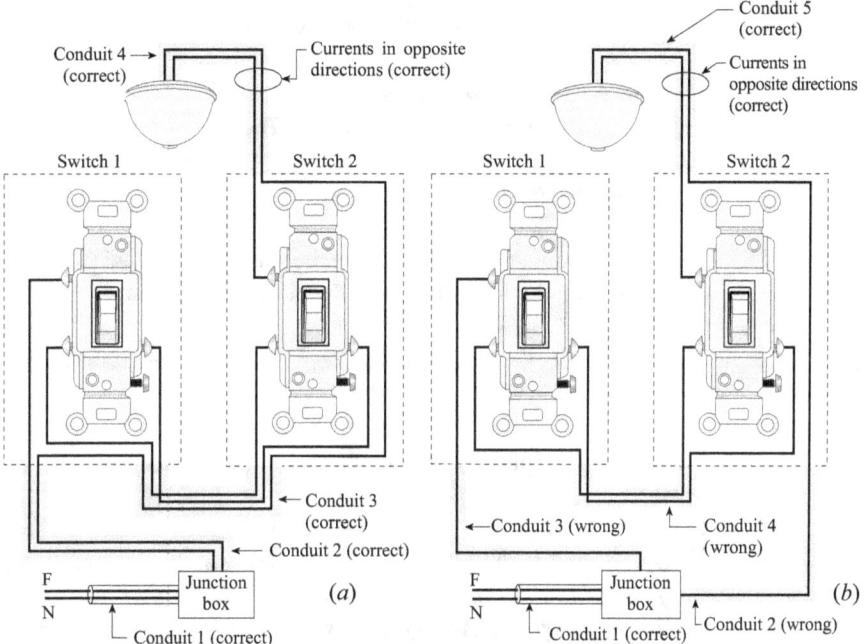

Fig. 75 (*a*) Proper wiring of two three-way switches to prevent induction heating in metal ducts. (*b*) Improper wiring causes induction heating.

As can be seen in **Fig. 75**(*a*), in each of the conduits 1, 2 and 4 there is a phase and a neutral, which carry current in opposite directions, which cancels the magnetic effect that causes induction heating. In conduit 3, although it contains three conductors, only two of them carry current at the same time, since of the conductors inside only one conducts when connecting the three-way switches with the lamp.

In the wiring of **Fig. 75**(*b*), conduits 2, 3 and 4 house a single current carrying conductor and will therefore produce induction heating in the metal raceways. conduits 1 and 5 do not present this problem, since they have conductors that carry currents in opposite directions. We have already said that in conduit 4, although it contains two conductors, only one of them carries current at the same time.

The switches must comply with the following:

Grounding conductor: Switches must not disconnect the grounding conductor of a circuit.

Exception: A switch shall be permitted to disconnect a grounded conductor when all circuit conductors are disconnected simultaneously or when the device is installed and the grounding conductor cannot be disconnected before all live circuit conductors have been disconnected. .

With regard to the use of switches in wet areas, the following should be considered:

Weather Protection: Surface mounted switches in wet locations must be enclosed in a weatherproof enclosure or box. They should not be installed in humid environments of bathtubs or showers. If they are mounted flush with a surface they will be equipped with a waterproof cover.

When, for some reason, blade switches must be installed (which is not so common for a residence), the following must be taken into account:

Placement: SPST switches will be installed in such a way that gravity does not tend to close them..

Regarding the placement height of the switches (see **Fig. 76**), the following is established:

Height: All switches must be positioned so that they can be operated from an easily accessible location. They must be installed so that the center of the activation levers, when in their highest position, are no more than 2 m (6 ft 2 in) above the floor or work platform.

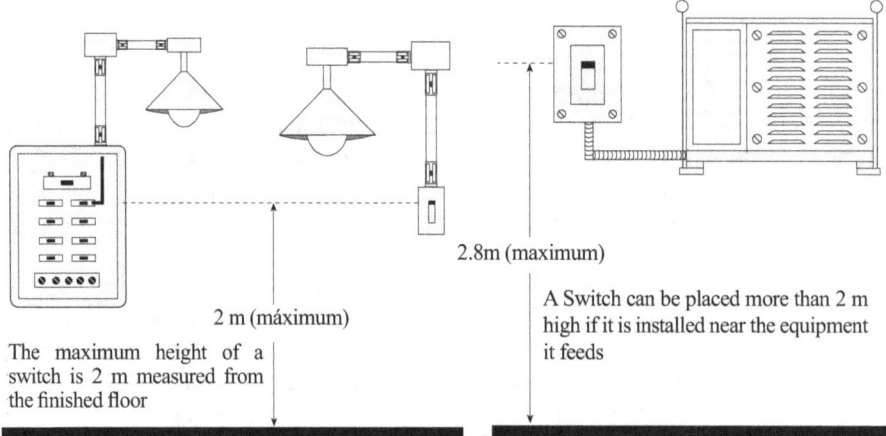

2.8m (maximum)

A Switch can be placed more than 2 m high if it is installed near the equipment it feeds

2 m (máximum)

The maximum height of a switch is 2 m measured from the finished floor

Fig. 76 Maximum heights of the switches.

Exception 2: *Switches installed adjacent to motors, appliances, or other equipment they feed, may be located at a height greater than 2 m and shall be accessible by portable means.*

> *Following what has been established in the previous paragraphs, it is common to place the switches at a height of 120 cm to 140 cm from the finished floor to the middle of the box that houses the switch.*

In relation to grounding, the following aspect shall be taken into account:

(1) *The switch is fastened with metallic screws to a metallic box or to a non-metallic box that has the means for grounding.*

(2) *An equipment grounding conductor connects to a grounding termination on the switch.*

Fig. 77 illustrates the two previous situations.

Regarding the details on the manufacture of the switches, it is important to consider the following: *the front metal covers of the switches will be made of ferrous metal with a thickness of not less than 0.76 mm, or of non-ferrous material with a thickness of not less than 1.02 mm. Non-metallic front covers of insulating material shall be non-combustible and not less than 2.54 mm thick, but shall be permitted to be less than 2.54 mm thick if reinforced to provide adequate mechanical strength.*

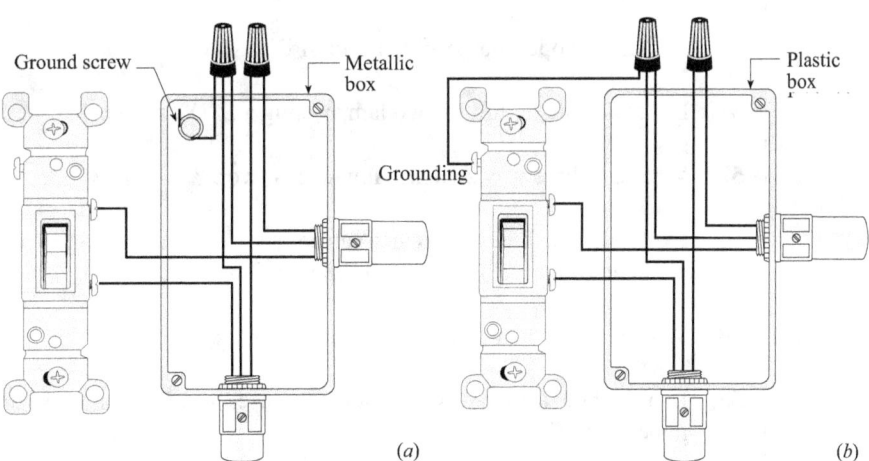

Fig. 77 (*a*) By fastening the switch, with screws to the metal box the grounding is ensured. (*b*) When the box is plastic, the grounding conductor is connected to the grounding screw of the switch.

20. ELECTRICAL SYMBOLS FOR SWITCHES

The symbols used to represent switches on electrical drawings are:

S : Single pole switch. S_3 : Three-way switch.

S_4 : Four way switch.

Think...
Explain...

47. What is an electrical switch?

48. Why is a switch said to be a binary element?

49. Draw the basic structure of a switch and explain its operation.

50. Explain the operation of SPST and SPDT switches.

51. Describe how to turn on two lamps using a SPDT switch.

52. What is a three way switch? How does it work?

53. Explain how a three-way switch can be used to turn a lamp, outlet, or electrical appliance on or off from two different points.

54. What is a four way switch? How does it work?

55. Explain how a four-way switch can be used to turn a lamp on or off from three or more points.

56. How are the electrical characteristics of a switch specified?

57 Explain all about the usage regimes for switches, including one-way, two-way, and three-way switches.

58. What is the typical current of a tungsten light bulb when hot or cold? How does this affect the selection of a switch?

59. What is induction heating in ferrous metal conduit? How does this heating occur?

60. Explain how three-way switch wiring can cause induction heating in metal ductwork. Can non-metallic ducts be induction heated?

61. Why should a switch not disconnect the grounding conductor?

62. Describe what the regulations state regarding the use of switches in wet areas.

63. Explain about the installation position of knife switches.

64. What do the regulations establish regarding the placement height of the switches?

65. What do the regulations determine in relation to the grounding of the switches?

66. What do the standards establish regarding the characteristics of the front covers of the switches?

67. When connecting three-way switches, to which terminals of the other switch should the traveling terminals of one of the switches be connected?

68. Is it always necessary to connect the equipment grounding conductor of a non-metallic sheathed cable to the ground screw of the switch box? Mention details about it.

69. How is the front cover of a switch grounded?

Ejercicios

In the following problems use the pictures of switches, lamps, wires, raceways, outlets, and boxes already shown in this book.

10. In the circuit of **Fig. 78** where the lamp is controlled by a switch, power enters through the switch. Draw the corresponding pictoric diagram .

11. In the circuit of **Fig. 79** where the lamp is controlled by a switch, power enters through the lamp. Draw the corresponding pictoric diagram.

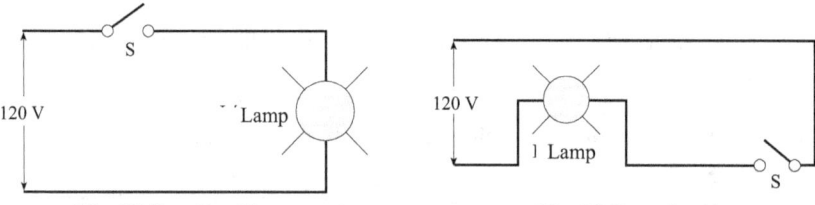

Fig. 78 Exercise 10   **Fig. 79** Exercise 11

12. La **Fig. 80** corresponds to the wiring of a lamp controlled by a simple switch. The inpuy power is directed without interruption to an outlet. Draw the wiring circuit diagram. The ground wire is omitted.

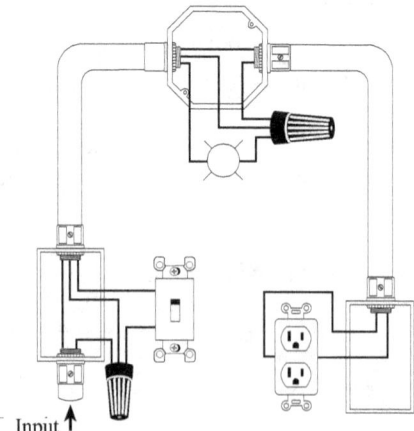

Fig. 80 Exercise 12

13. From the electrical circuit in **Fig. 81**, where three-way switches are used, draw the corresponding wiring.

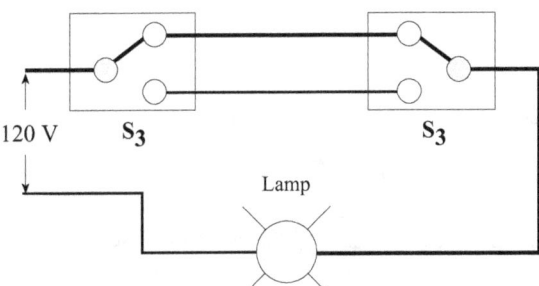

120 V S_3 S_3

Lamp

Fig. 81 Exercise 13

14. Draw the elementary diagram of the electrical installation whose wiring is shown in **Fig. 82**. The 120 V supply enters through the lamp.

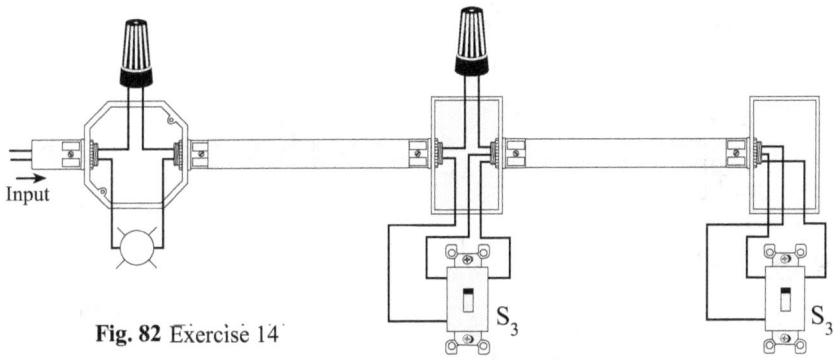

Input

Fig. 82 Exercise 14 S_3 S_3

15. A lamp is controlled from three different points, as indicated in **Fig. 83**. From the wiring diagram shown draw the elementary diagram. Power enters through one of the three-way switches.

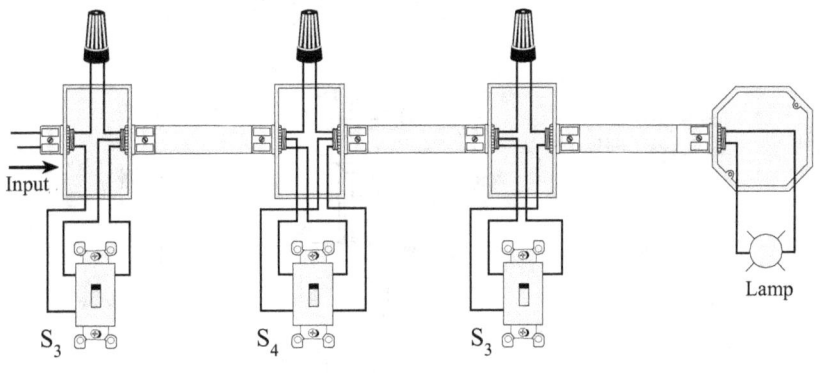

Input

Lamp

S_3 S_4 S_3

Fig. 83 Exercise 15.

16. Some switches use a pilot light to indicate when a lamp is on or off. This light is obtained from a neon tube in series with a resistor. Explain how this system works, referring to **Fig. 84.17.**

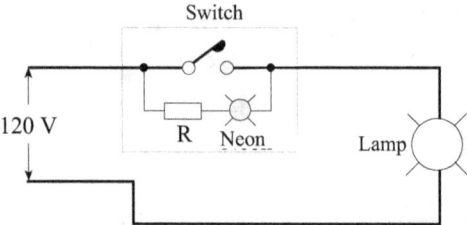

Fig. 84 Exercise 16.

17. Explain how the circuit whose wiring is shown in **Fig. 85** works. Draw the elementary circuit diagram.

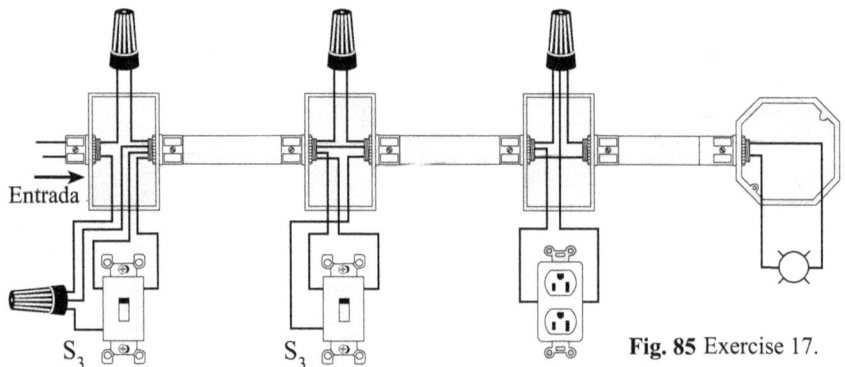

Fig. 85 Exercise 17.

18. Explain how the circuit whose wiring is shown in **Fig. 86** works. Draw the circuit diagram.

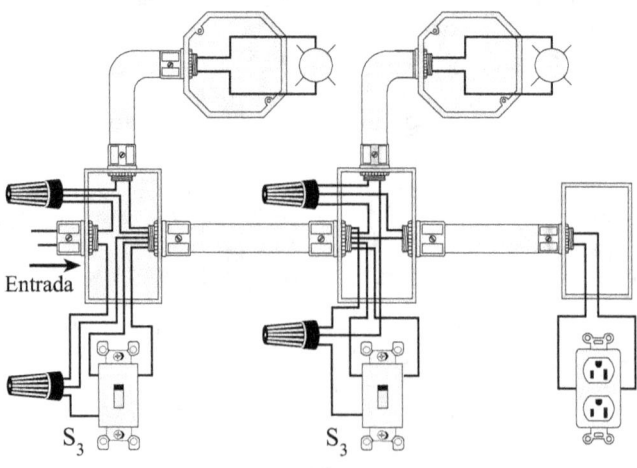

Fig. 86 Exercise 18.

19. Draw the wiring diagram for **Fig. 87** so that the lamps can be controlled by the switches. The outlet must be also energized. Use the proper conductors, boxes, and connectors.

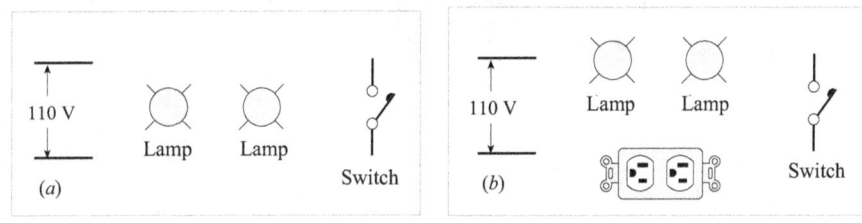

Fig. 87 Exercise 19.

20. Complete the connections for the control of the two lamps in **Fig. 88** from two different points and by using two three-way switches.

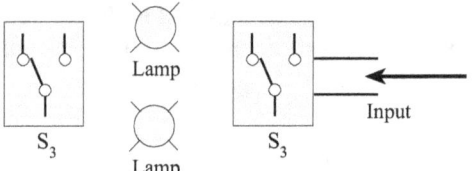

Fig. 88 Ejercicio 20.

21. Draw the wiring diagram for **Fig. 89** so that the lamps can be controlled by the switches. The outlet must be also energized. Use the proper conductors, boxes and connectors.

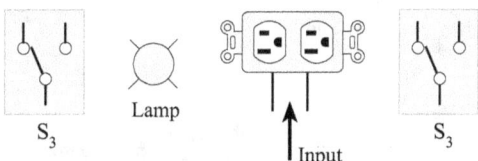

Fig. 89 Exercise 21.

22. Complete the connections in **Fig. 90** and draw the wiring diagram for control of the lamp from three different points, using three-way and four-way switches. The 110V power line enters through the lamp.

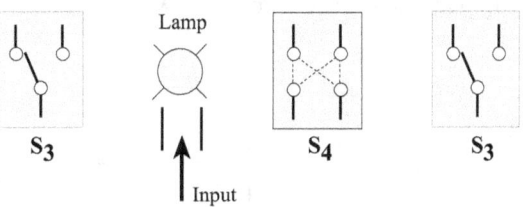

Fig. 90 Exercise 22.

23. Complete the connections in **Fig. 91** and draw the wiring diagram to control the lamp from three different points using three-way and four-way switches. 110V power enters through the 4-way switch.

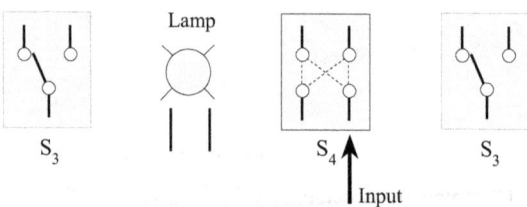

Fig. 91 Exercise 23.

24. Complete the connections in **Fig. 92** and draw the wiring diagrams for control of the lamp from three different points, using three-way and four-way switches.

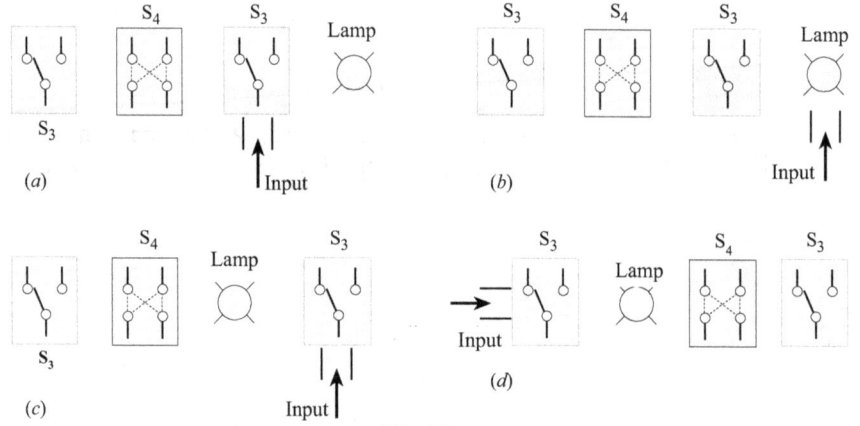

Fig. 92 Exercise 24.

25. Using **Fig. 93** explain how the lamp can be controlled from four different points and by using 2 three-way and 2 four-way switches.

26. Explains how the circuit in **Fig. 94** works.

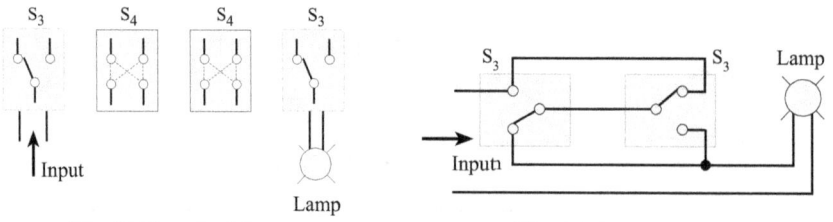

Fig. 93 Exercise 25.　　　　**Fig. 94** Exercise 26.

LOCATION OF OUTLETS AND LUMINAIRES

21. THE ELECTRICAL PROJECT. GENERALITIES.

The planning and project of an electrical installation is the result of extensive consultation which includes the owner of the building, the architect of the work and the
electrical engineer who will develop the electrical project. The latter must take into
account the habits and requirements of those who will occupy the building, as well as
the appliances and electrical equipment necessary to guarantee the comfort and quality
of life of the users. To do this, the engineer will strictly observe the requirements of a
good electrical installation: safety, capacity, accessibility, flexibility and economy. This
book emphasizes the location of outlets and luminaires in residential units.

Often, in countries with significant poverty rates, there is a tendency to sacrifice the
above requirements in order to obtain a drastic reduction in costs. As the fundamental
objective is to ensure the safety of people who interact with electricity, there is no need
to establish differences between social housing, rural housing, housing for the middle
class or housing for the upper class. In this sense, the security of the human being is
valued, regardless of their social status. This tendency to endanger the lives of those
who have the least must be reversed in practice, designing electrical systems that guarantee the safety of all those who are exposed to the dangers of electricity on a daily
basis. What should differentiate the electrical design is the more or less extensive use
of those equipment and appliances that are present in households of different income
levels. The electrical engineer has to face this challenge, reflecting in his design how
important safety is for any user of electrical energy without economic barriers inter-
vening to endanger human life.

A good electrical project adheres, at least, to the following characteristics:

1. The use of approved quality materials for installation. This includes the materials used in the manufacture of conductors, conduits, boxes, outlets, switches,
   lighting fixtures, etc.

2. Sufficient number of outlets, points of light and switches located in those places
   that facilitate the use of the electrical installation.

3. Boards with the capacity to respond to future expansions.

4. Additional conduis for possible expansions of the electrical installation.

5. Connection capable of supporting the present and future design load.

6. Use of a grounding conductor in the installation and grounding of the metal covers of equipment to avoid electric shocks.

7. Use of ground fault interrupters (GFCI) and arc fault interrupters (AFCI) to disconnect circuits whose live conductor is grounded or which produce sparks capable of causing a fire.

Along with its consumption in watts it is worth making a list of the equipment that is usually found in the various rooms of a residence. Although not all of them may be found in a specific electrical installation, mentioning them can serve as a guide for the design of residential branch circuits. The consumption of the artifacts may differ from those shown in **Table 4***.

Electrical Equipment/Appliance	Consumption (W)	Equipo Eléctrico/Artefacto	Consumption (W)
Air conditioner 24000 BTU (2 tons)	1900	Kitchen (4 burners)	8000
Air conditioner 30000 BTU (2.5 tons)	2800	Kitchen (oven + 4 burners)	12000
Air conditioner 36000 BTU (3 tons)	2900	Computer	60–250
Air conditioner 60000 BTU (5 tons)	4900	Freezer (14 cubic feet)	350
Air conditioner split 9000 BTU	820	Electric knife	360
Air conditioner split 12000 BTU	1260	Portable dehumidifier	90
Air conditioner split 15000 BTU	1410	Electric shower	3500
Air conditioner split 18000 BTU	1840	Stereo equipment	100
Air conditioner split 24000 BTU	2300	Bottle sterilizer	500
Air conditioner split 36000 BTU	2660	Large oven	4000–8000
Air conditioner.(window)12000 BTU	800	Humidifier	40
Air conditioner.(window) 15000 BTU	1410	Deskjet printer	20
Air conditioner.(window) 18000 BTU	1840	Laser printer	400
Air conditioner.(window) 24000 BTU	2300	Automatic washing-machine	500
Can opener	120	Diskwasher	1200–1500
Vacumm cleaner	650	Blender	300
Blender	200	Razor	20
Water pump 1.5 HP	1120	Sewing machine	100
Water pump 1/3 HP	250	Microwave	600–1500

Table 4 Equipment and appliances commonly used in a residence and their typical consumption in watts.

---

* The consumptions shown are only illustrative. Actual consumption may vary according to technological advances.

Electrical Equipment/Appliance	Consumption (W)	Electrical Equipment/Appliance	Consumption (W)
Floor polisher	300	1/2 in. drill	750
Radio	20–70	1/4 in. drill	250
Refrigerator	400	19 inch TV	200
Sandwich Maker	650	25 inch TV	250
Electric pan	1300	Bread toaster	800–1500
Hair dryer	1875	Tamal toster	1200
Clothes dryer (120V)	1600	Garbage disposal	10–50
Clothes dryer (220V)	5000	Ceiling fan	10–50
1 in. drill	1000	Portable table fan	10–25

Table 4 *Continuation.* Equipment and appliances commonly used in a residence and their typical consumption in watts.

To optimize the electrical installation in a residence it is recommended that the project design include the following steps:

1. Ensure that the power supply is available.

2. Establish conversations with the owner of the residence, if it is an individual development, or with the group of families, if it is a matter of collective projects or of social interest, in order to specify the equipment and artifacts to be used in the home. Take forecasts for future expansions.

3. Specify where the different points of outlets, lamps, switches and all those outlets necessary to develop the wiring of the installation in the architectural plans are going to be placed. This should be done in close collaboration with the architect and the residential owners. Here it is necessary to distinguish between outlets for general use and those that will power individual appliances and equipment.

4. Locate the placement sites for the main board and other boards.

5. Calculate the number of lighting circuits and outlets. Add reservation circuits for future extensions.

6. Draw the wiring of the lighting circuits and outlets, as well as establishing the way to connect them to the main board and other boards.

7. Calculate the size of the conductors and ducts of the electrical installation.

8. Determine the protections of each of the circuits of the electrical installation.

9. Verify that the voltage drop does not exceed what is suggested by the codes.

10. Select the type of connection: aerial or underground.

11. Calculate the size of the service conductor.

12. Design communication and signal systems.

In this chapter we will concentrate on the third of the above steps. For this we will consider the different environments of the residences and the location of the outlets, light points and switches that make up the installation.

22. GENERALITIES ABOUT THE LOCATION OF OUTLETS, LAMPS AND SWITCHES IN THE DIFFERENT ENVIRONMENTS OF A RESIDENCE

Before addressing the specific objective of this section it is convenient to make some general observations that optimize the design of the electrical installation:

1. When there are adjoining rooms considerable savings will be obtained by placing two outlets opposite each other, as indicated in **Fig. 95**.

2. It is convenient, when possible, to place one of the outlets below the switch that controls the ambient light. This provision prevents this outlet from being hidden behind any furniture, since it is unlikely that any furniture will be placed under the switch (see **Fig. 96**).

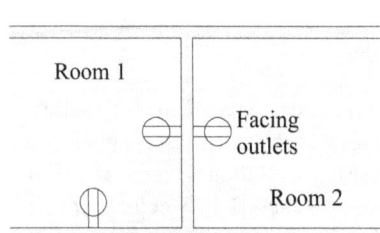

Fig. 95 Placement of outlets in contiguous spaces.

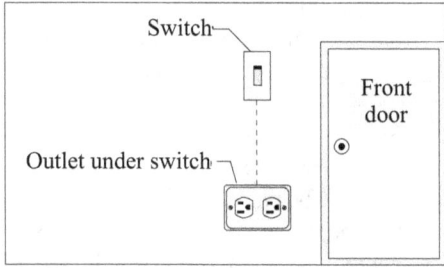

Fig. 96 In order to prevent the outlet from being hidden by a piece of furniture, it is convenient to place it under the switch that controls the ambient light.

3. Aun cuando no está expresamente prohibido por las normas, los interruptores de lámparas no deben colocarse detrás de las puertas de los distintos ambientes. Esto facilita el encendido o apagado de las lámparas. Ver **Fig. 97**.

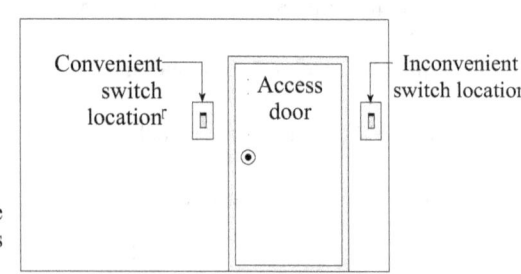

Fig. 97 It is not advisable to place the lamp switches behind the access doors.

4. Despite the fact that maximum distances between two outlets have been established, a good design must provide for the inclusion of enough outlets below these distances, avoiding that they become limiting rules. Likewise, the exchange of information with the owner and the architect or civil engineer must define where the furniture will be placed within specific environments, so that the furniture does not hide the electrical outlets.

5. Placement of general purpose electrical outlets: electrical outlets must be installed in the different rooms of a residence (kitchen, dining room, living room, library, bedrooms, hallways, etc.), so that no point, measured horizontally along the floor line, in any wall space, is more than 6 feet (1.80 m) from an electrical outlet. This means that the maximum distance between one outlet and another cannot be greater than 3.60 m, as indicated in **Fig. 98**. This rule does not apply to bathrooms, laundry rooms or garages, environments where there are specific situations from the electrical installation point of view. Note that *a wall space is defined as one with a length equal to or greater than 60 cm (2 feet), including the distance around corners, not interrupted* along *the floor line by doors, chimneys, or other similar openings.*

In **Fig. 99** it can be seen that between the sockets A-B, B-C and C-D the distances are 3.6 m, while the distance D-A is 2.2 m, a length less than the maximum distance allowed. One way to start locating outlets in a room is to measure 6' from both ends of the front door, then take 12' wall intervals to place the other outlets, as shown in **Fig. 99**. General outlets will be placed at a height of 30 cm above the finished floor. It is worth mentioning that the placement of the electrical outlets in the bedroom in **figures 98** and **99** only tries to underline the distances between them, without taking into account the

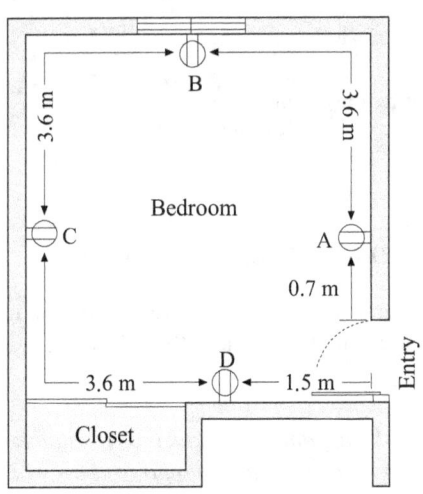

Fig. 98 Between one outlet and another the distance, measured along the wall, should not be greater than 3.6 m (12 ft).

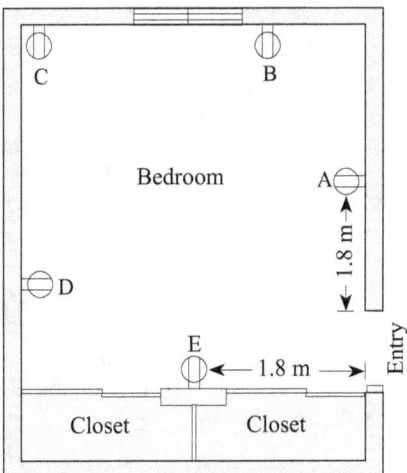

Fig. 99 A good practice to start the location of the electrical outlets is to place the first two at a distance of 1.8 m, on both sides of the entrance door opening, and, from there, locate the rest of the electrical outlets .

convenience of their location at certain points. The most convenient location for power outlets should take into consideration the placement of furniture within the room and will be discussed later in this chapter.

6. In any wall space with a length of 60 cm or more, an electrical outlet must be placed. See **Fig. 100**.

7. Floor outlets, placed less than 45 cm from the wall, must be considered for the purposes of the 3.6 m distance discussed in the previous points.

8. In corridors equal to or greater than 3 m in length, at least one electrical outlet must be installed. For the purposes of measuring this distance, the median line of the corridor will be taken into account, as indicated in **Fig. 101**.

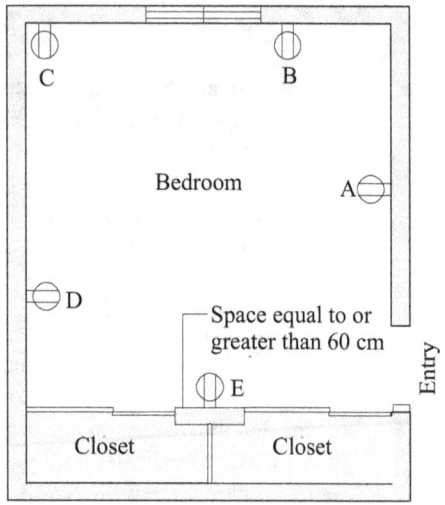

Fig. 100 An electrical outlet must be placed in any wall space with a length equal to or greater than 60 cm.

9. In order to ensure proper maintenance of heating, air conditioning or refrigeration equipment, an electrical outlet must be installed at a distance of not less than 7.5 m from such equipment.

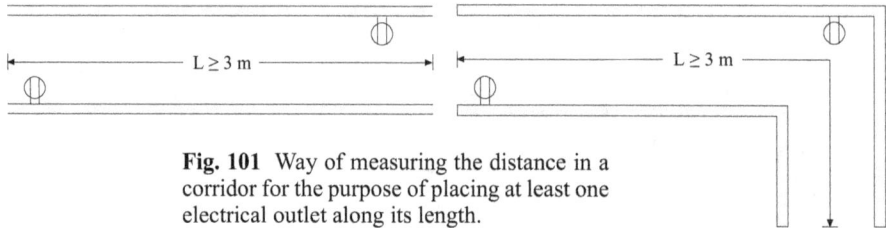

Fig. 101 Way of measuring the distance in a corridor for the purpose of placing at least one electrical outlet along its length.

23. OUTLETS IN THE KITCHEN AND IN THE DINING ROOM

The kitchen is one of the areas of a residence that requires more attention in terms of electrical installation. A good electrical design will consider the following in relation to the kitchen:

Circuits for small appliances: *In the kitchen, pantry, breakfast room, dining room, or similar areas, two or more small appliance circuits should feed all wall and floor outlets, all outlets on countertops and all refrigeration outlets.*

Exception: The refrigeration equipment outlet may be powered from a single circuit of 15A or greater.

Let's interpret the above rule. From this it can be deduced that for the kitchen at least two 20 A circuits must be allocated for small appliances (blender, toaster, etc.), which will feed the outlets that go on the walls and floors of the rooms described (kitchen, pantry, breakfast room, dining room, etc.), to the sockets located above the kitchen cabinet and to the sockets of the refrigeration equipment (fridge and freezer).

On the other hand, the mentioned exception establishes that the refrigerator can have an outlet connected to an individual circuit. This is planned to prevent voltage fluctuations, when the refrigeration equipment starts, from affecting other branch circuits in the installation.

As mentioned above (point 5), general kitchen outlets should be installed so that no point along the floor line on an uninterrupted wall is more than 6 feet from an outlet. *The two or more small appliance circuits must not power appliances such as automatic dishwashers, electric stoves, garbage disposals, trash compactors, microwave ovens, or outside kitchen outlets.* Exceptions are sockets for electric clocks and those used to light gas and electric stoves and ovens, which can be connected to the circuits of small appliances.

The placement of electrical outlets on kitchen cabinets is also regulated:

Distance: an electrical outlet will be installed in any space above the kitchen cabinets that have a length equal to or greater than 30 cm. Outlets must be installed so that no point along the line is more than 24 inches (60 cm) measured horizontally from an outlet in that wall space.

*Exception: an outlet on a wall directly behind a stove or sink and that is 12 inches or less in length is not required (see **Fig. 103**). The space behind those artifacts does not count for the purposes of the distance described above.*

According to the above, the distance between two consecutive power outlets must not exceed 1.20 m (48 inches). Note that we are talking about a wall space, as defined above. Therefore, the spaces occupied by kitchen equipment, sinks (punch bowls), dishwashers and similar equipment are not taken into account when measuring the distance between outlets.

Kitchen spaces between which are kitchen counters, refrigerator, freezer and sink (punch bowls) must be considered as separate spaces.

All outlets placed above kitchen cabinets must be of the GFCI type. On the other hand, outlets behind electrical equipment such as refrigerators, freezers, and automatic dishwashers do not need to be GFCI protected.

On each side of appliances such as automatic dishwashers, stoves, and sinks, a distance less than or equal to 60 cm will be left between the appliance and the electrical outlet. This is a criteria for starting to lay out outlets on cabinet tops.

In **Fig. 102** the distribution of electrical outlets in the kitchen room is presented considering the criteria described above. In such a figure it can be seen that only the outlets that are visible on the top of the kitchen cabinet and corresponding to small appliances (identified as T_{6PA}, T_{7PA}, T_{8PA}, T_{9PA}, T_{10PA} and T_{11PA}) are of the GFCI type. Any outlet behind the range that is used for the electronic ignition, range hood lights, or any other use does not require GFCI protection.

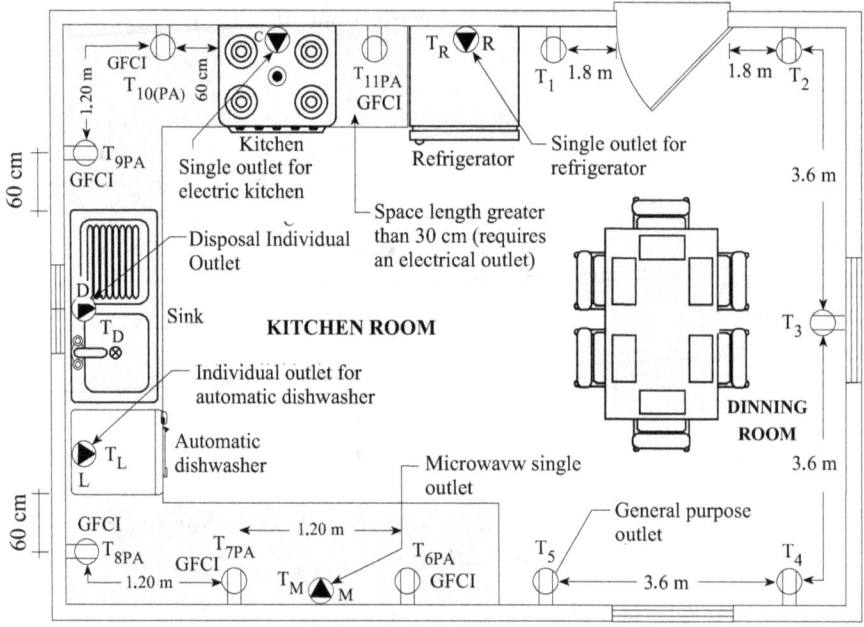

Fig. 102 Distribution of outlets in the kitchen room, as established by electrical codes.

The T_{8PA} and T_{9PA} sockets are placed 60 cm away, next to the dishwasher and the sink. The T_{10PA} outlet is 60 cm from the cooker. Between the cooker and the refrigerator there is a work surface with a width greater than 30 cm and therefore requires the presence of a power outlet (T_{11PA}). Note that the distance between the T_{9PA} and T_{10PA} outlets can be equal to or less than 1.20 m. We have assumed that it is 1.20 m, but it could be less. Between the sockets (T_{6PA}– T_{7PA}) and (T_{7PA}– T_{8PA}) there are 1.20 m.

In the kitchen room there are five individual electrical outlets (T_M, T_L, T_D, T_C and T_R), which correspond to the microwave, automatic dishwasher, garbage disposal, electric stove and refrigerator, respectively.

Five general use outlets (T_1, T_2, T_3, T_4, and T_5) serve to feed the dining room, separated by 3.6 m. To make the distribution in this area, we started with two electrical outlets

$(T_1$ and $T_2)$ placed 1.8 m from the ends of the access door to the dining room.

Fig. 103 clarifies what was contemplated in the last mentioned exception: when the distance X does not exceed 30 cm, it is not necessary to install an electrical outlet behind the appliance.

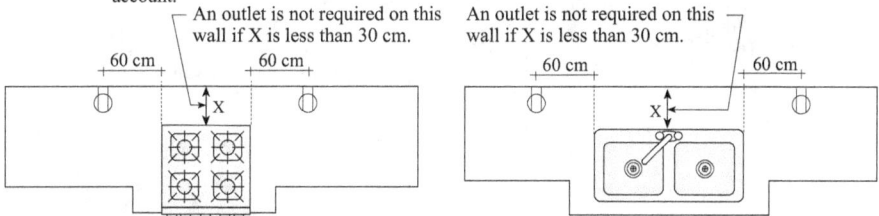

Fig. 103(*a*) Placement of outlets on the wall of kitchen cabinets: It is not necessary to place an outlet behind a stove or sink, whose installation is as shown in the figure, when the distance X is not more than 30 cm.

- -

No point along the line joining two power outlets shall be more than 60 cm from any one of them. If X is less than 30 cm, the horizontal distance behind the artifact will not be taken into account.

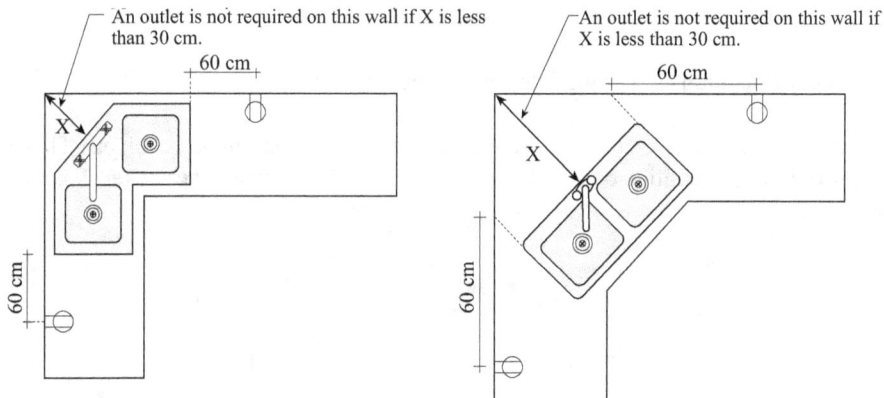

Fig. 103(*b*) Placement of outlets on the wall of kitchen cabinets: It is not necessary to place an outlet behind a stove or sink, whose installation is as shown in the figure, when the distance X does not exceed 30 cm.

Another situation to consider in a kitchen room has to do with isolated furniture that is not an integral part of the kitchen cabinets. These isolated pieces of furniture are known as peninsulas and islands and have as a distinctive characteristic the fact that they do not have walls behind or in front of them, as indicated in **Fig. 104**. They must be fed in accordance with the provisions of the rules that a we consider below.

At least one electrical outlet will be installed in each peninsular space that has a minimum length of 60 cm or a minimum width of 30 cm. The peninsular space is measured from the side that connects it with the rest of the kitchen cabinet. On peninsulas and islands, electrical outlets will be installed at a distance no greater than 30 cm below their top.

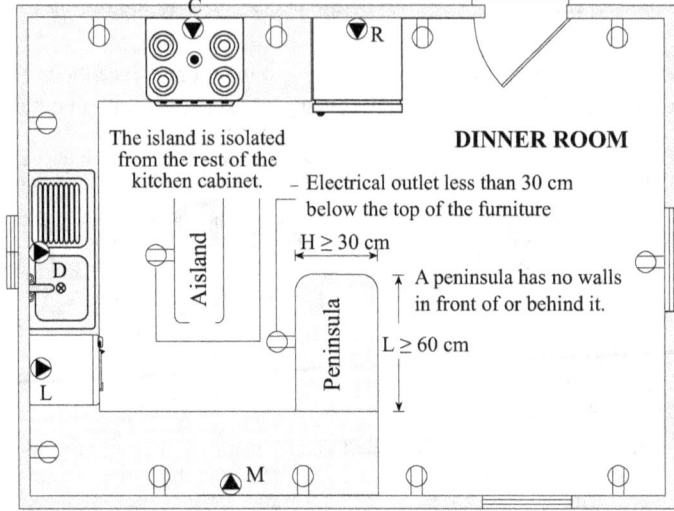

Fig. 104 Kitchen room with peninsula and island. These are distinguished by not having walls behind them.

As for the islands, an electrical outlet must be installed on each island that has a minimum length of 60 cm or a minimum width of 30 cm.

Other considerations regarding kitchen electrical outlets include:

1. Electrical outlets are not allowed to be placed face up on kitchen cabinets. This is to prevent the penetration of liquids and other substances through the socket slots. See **Fig. 105**.

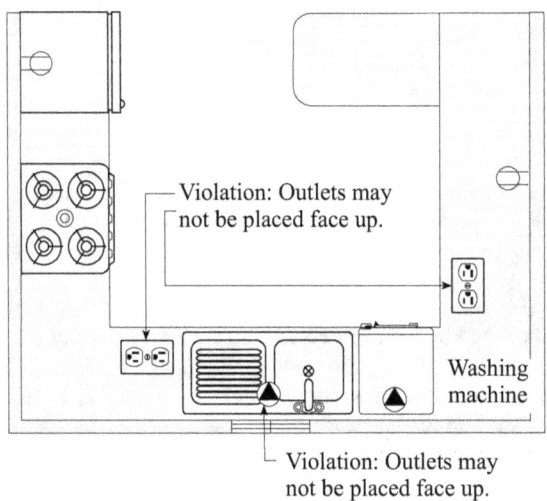

Fig. 105 Installation of outlets face up in kitchen cabinets is not permitted.

2. Power outlets should not be placed higher than 50 cm above kitchen countertops. Watch **Fig. 106**.

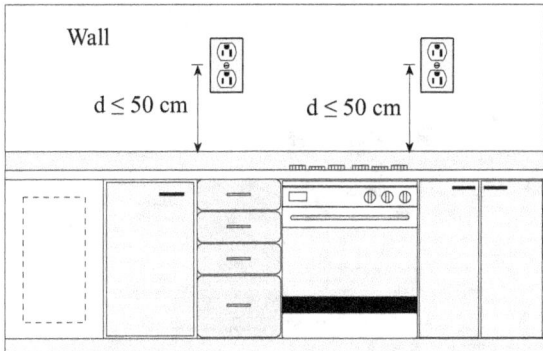

Fig. 106 The installation of outlets at a height less than 50 cm from the kitchen counters is not allowed.

3. Outlets that are not easily accessible, such as those behind a refrigerator, automatic dishwasher, or garbage disposal, are not considered part of the outlets to be placed above kitchen cabinets and do not require to be of the GFCI type. Watch **Fig. 107**.

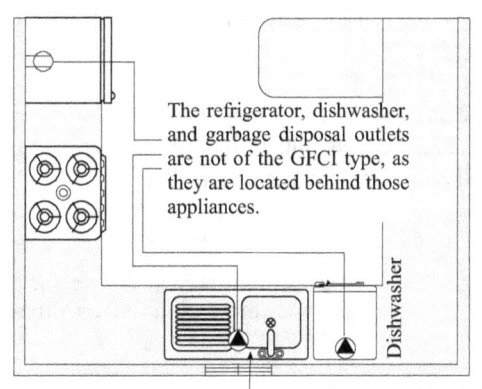

The refrigerator, dishwasher, and garbage disposal outlets are not of the GFCI type, as they are located behind those appliances.

Fig. 107 GFCI outlets are not required for the refrigerator, dishwasher, and garbage disposal because they are not countertop outlets.

Disposal outlet (under sink)

4. Outlets must not be installed under a kitchen counter, island or peninsula that extends more than 15 cm above its base cabinet. See **Fig. 108**.

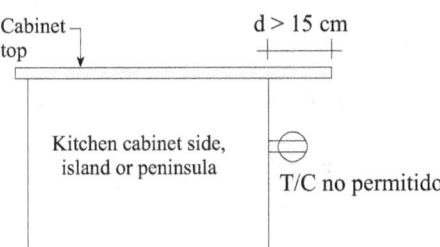

Fig. 108 Installation of outlets under the top of a kitchen cabinet, island or peninsula is not permitted when it extends more than 15 cm above its base cabinet. When the top protrudes less than 15 cm, it is allowed to install outlets at a distance of 30 cm from the top.

5. Every microwave oven must be connected to an individual electrical outlet. The electric cooker must also have its own individual electrical outlet.. **Fig. 109**.

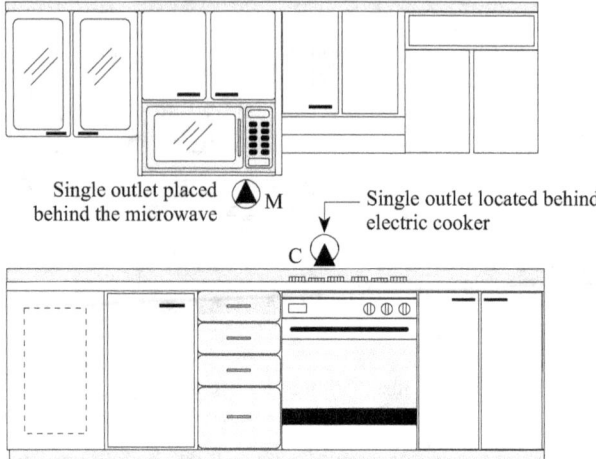

Single outlet placed behind the microwave

Single outlet located behind electric cooker

Fig. 109 Both the microwave oven and the electric stove must have individual outlets.

6. On a peninsula or island, outlets may be installed above the top level by using the appropriate framing. **Fig. 110**.

7. It is convenient to leave an outlet for an electrical outlet under the sink in the kitchen room for the installation of an electric water filter.

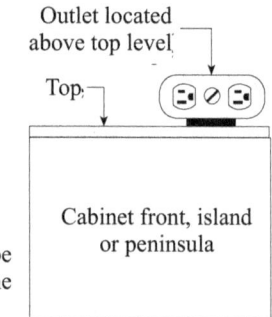

Outlet located above top level

Top

Cabinet front, island or peninsula

Fig. 110 Outlets on kitchen cabinet tops should be positioned so that their faces protrude above the cabinet top..

In relation to lighting in the kitchen room, it is recommended to leave outlets for lamps in the places described below. Likewise, it is advisable to use LED lamps in all environments. This will result in significant energy savings.

a) Dining room and kitchen room: general ceiling lights that allow lighting the central area.

b) Dishwasher (sink): focused light to facilitate the task in this area.

c) Cabinets: lights placed on those cabinet surfaces not adequately covered by general lighting.

d) Eating areas: attention should be paid to places in the kitchen room that serve as eventual eating areas, such as breakfast or dinner areas. These can be located on the same kitchen cabinet and must be adequately illuminated.

e) In some cases, it may be necessary to illuminate the interior of the floor cabinets in the kitchen. It is important to discuss this with the owner, builder or architect.

The kitchen, when it has the hood to collect the gases released from the pans and pots, normally has its own lighting.

For the environments of the kitchen room and the dining room of **Fig. 102**, a distribution of luminaires and their respective control switches can be made as the one presented in **Fig. 111**.

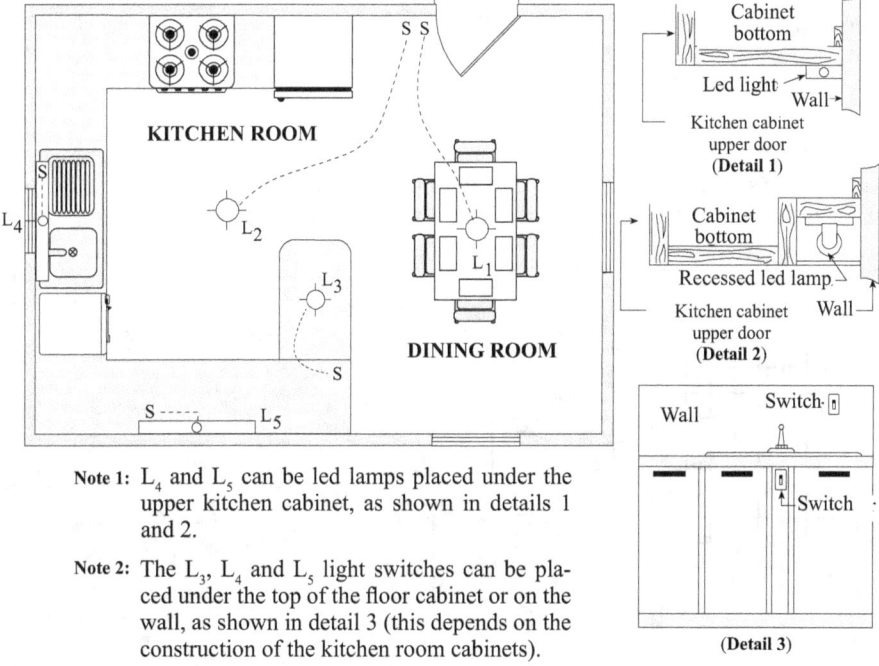

Note 1: L_4 and L_5 can be led lamps placed under the upper kitchen cabinet, as shown in details 1 and 2.

Note 2: The L_3, L_4 and L_5 light switches can be placed under the top of the floor cabinet or on the wall, as shown in detail 3 (this depends on the construction of the kitchen room cabinets).

Fig. 111 Distribution of lamps and control switches in the dining room and kitchen room.

Lamps L_1 and L_2 correspond to the general lighting of the kitchen and dining room. They are controlled by switches placed on the moving side of the door.

In the upper part of the peninsula, a place that can eventually be used for eating, a ceiling light (L_3) has been provided, controlled by a switch located below the top of the cabinet (see **Detail 3**). In order to illuminate the area where the sink is located, a led lamp (L_4) is provided above it, controlled by a switch that can be placed or below the top of the cabinet, if the kitchen furniture allows it, or in the part of the wall that is above the cabinet, as indicated in **Detail 3** of **Fig. 111**.

Se instalará una lámpara fluorescente (L_5) en la parte más larga del gabinete de cocina. Se controlará según el Detalle 3 de la **Fig. 111** o con un suiche ubicado convenientemente.

Las lámparas L_4 y L_5 se pueden instalar como lo muestran los detalles 1 y 2 de la **Fig. 111**. En el primer caso se colocan superficialmente, mientras que en el segundo se empotran entre la pared y la parte trasera del mueble superior del gabinete de cocina.

24. SALIDAS ELÉCTRICAS EN LA SALA DE BAÑO

A bathroom is defined as an area that includes a sink with one or more of the following additional items: sink, tub, and shower. The following provisions apply to bathrooms:

1. At least one outlet must be installed.

2. Outlets must be GFCI type.

3. Electrical outlets must be installed at a distance of less than 90 cm from the outer sides of the sinks, either on the back wall or on any wall that serves as a partition between the sink and the shower or sink. **Fig. 112**.

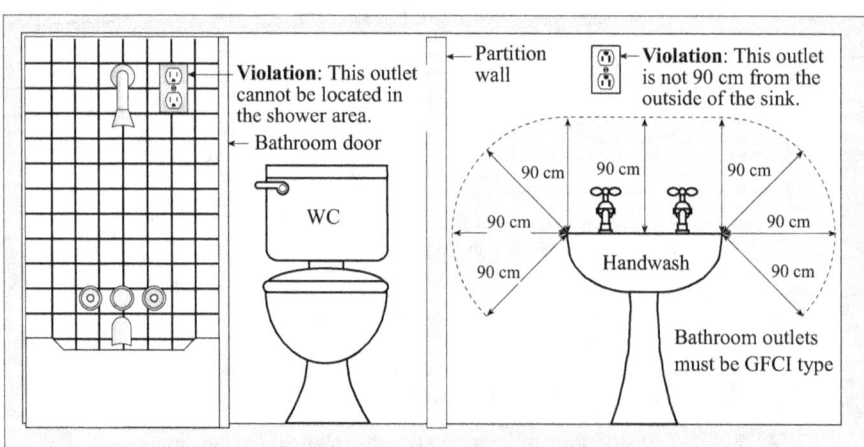

Fig. 112 In the bathroom, an outlet must be installed at a distance of less than 90 cm from the sink.

The bathroom is one of the places that must be designed based on safety criteria, since it is a place where water is combined with the presence of electrical devices. Hence the importance of using GFCIs in this environment.

4. If the sink is built into a piece of furniture it can be placed in front of the piece of furniture, at a distance not exceeding 30 cm below the top of the piece of furniture. **Fig. 113**.

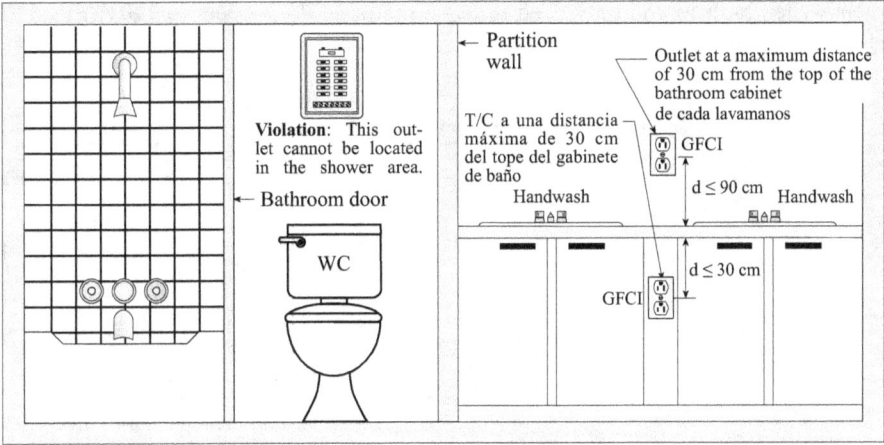

Fig. 113 When the sink is recessed, an electrical outlet may be placed less than 12 inches below the countertop.

5. Outlets must be powered from an individual branch circuit and must not supply power to other loads, such as lamp outlets. That same circuit can feed another bathroom, although it is not recommended due to the use it is given to power high-consumption hair dryers. It is recommended to use individual branch circuits for each bathroom.

6. Outlets may not be placed face up in bathroom cabinets. **Fig. 114**.

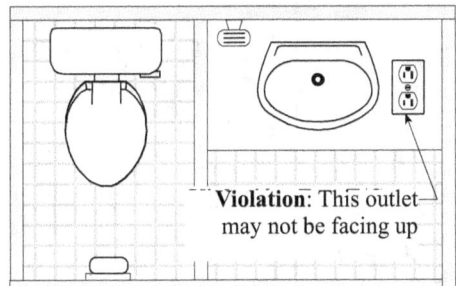

Fig. 114 In a bathroom, an outlet should not be placed with its front face up.

7. Outlets, regular or waterproof, are not allowed to be installed inside the shower or bathtub area. **Fig. 112**.

8. Electrical panels should not be installed in bathrooms. **Fig. 113**.

Regarding the lighting in the bathroom the regulations establish that there must be at least one lighting outlet controlled by a switch. General purpose luminaires are placed for the entire bathroom and lamps above the mirror. The latter are located in front of it or on the wall behind it. Some bathroom cabinets already have the light fixtures and their switch built in and only need to leave the outlet to connect to 120V. **Fig. 115** shows the lighting outlets for a typical bathroom.

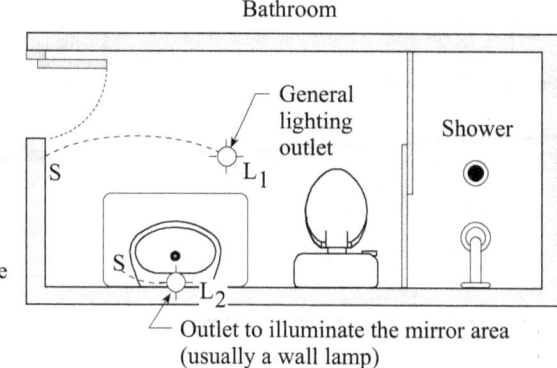

Fig. 115 Lighting outlets in the bathroom.

The placement of luminaires suspended by cord, track lighting, sconces or suspended ceiling fans is prohibited in an area between 90 cm horizontally and 2.5 m vertically from the top edge of a bathtub or shower cubicle, as indicated in the **Fig. 116**. However, luminaires, surface mounted or recessed, may be located within the restriction zone.

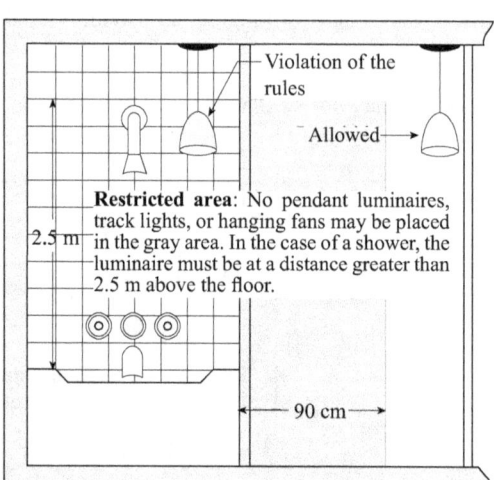

Fig. 116 The placement of pendant light fixtures, track lighting, and pendant fans is restricted on the shaded area.

It is worth mentioning that the luminaire used to illuminate the bathroom mirror should be placed in a way that facilitates the activities that take place in front of it. If it is positioned so that the light falls directly on the person's head, it will cause shadows on the face that will not allow a good image when shaving, combing, etc. In **Fig. 117** the incorrect and correct positions of locating the luminaire to achieve a good effect are shown.

Another variant of ceiling or recessed lighting in the mirror consists of placing several light bulbs on both sides of the mirror in addition to the lighting above it.

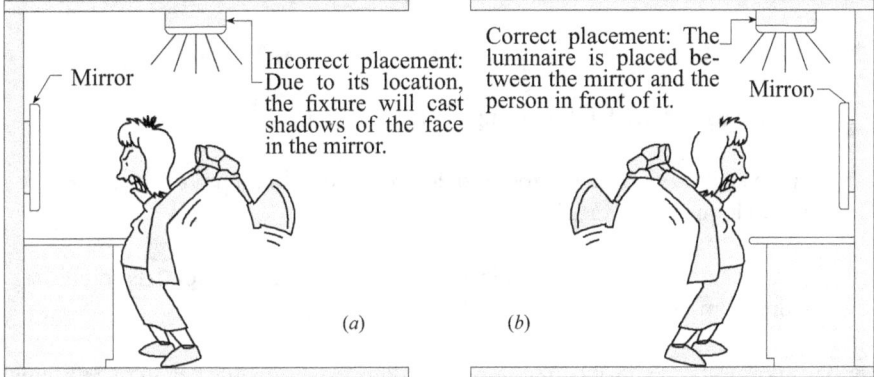

Fig. 117 Incorrect (a) and correct (b) positions of locating a luminaire in front of the bathroom mirror.

Finally, we must mention that the installation of switches in humid places, such as bathtubs and showers, is prohibited, unless they are installed as an integral part of the set that comes with the bathtub or shower.

25. ELECTRICAL OUTLETS IN BEDROOMS

The various activities carried out in the bedrooms require an adequate and versatile electrical installation. The different positions that the furniture can have inside the sleeping rooms determine the distribution of lights, outlets and switches.

We have mentioned that in a dwelling no point, measured horizontally along the floor line, in any wall space, should be more than 1.8 m from an electrical outlet. This is also valid for the rooms of a residence. As noted above, the latter implies that the maximum distance between outlets is 3.6 m. In other words, electrical outlets can be placed at distances of less than 3.6 m from each other, which allows you to play a little with the possible location of the beds and other furniture typical of the bedrooms. In order to avoid future remodeling of the electrical installations in the rooms, it is advisable to distribute the electrical outlets in such a way that they are not hidden behind the beds when, for any reason, their position is changed.

Among the considerations to take into account in the design of the electrical installation in a bedroom we can mention the following:

1. It is convenient to install three-way switches to turn on or off the general lighting lamp from the entrance of the room and from the bed.

2. An outlet must be provided to connect the television and any video playback equipment.

3. Study the possibility of lighting the interior of the cabinets.

4. Provide the outlet for a window or split type air conditioner.

5. Provide for the output of a ceiling fan.

6. Study the use of Arc Fault Interrupters (AFCI).

7. Include switches inside the room to control external lighting of the house, in order to improve its security.

8. It is convenient, although not mandatory, to use split-phase outlets to ensure continuity of service in the event of a phase failure.

Fig. 118 shows a bedroom and alternative locations for the double bed in two different positions.

In the first case, the bed is attached to the wall opposite the bathroom and we could think of the following options regarding the placement of the electrical outlets: 1) the T/C to connect a TV could be placed on the bathroom wall; 2) The outlet of the window air conditioner, or the console of a split unit, could be placed next to the window.

In the second case the double bed is placed in front of the window and we would have these options: 1) the socket for the TV could be placed in the closet, for which the design of the closet must be appropriate; 2) the outlet for the window air conditioner or the console of a split unit could be placed on the wall to the right side of the bed.

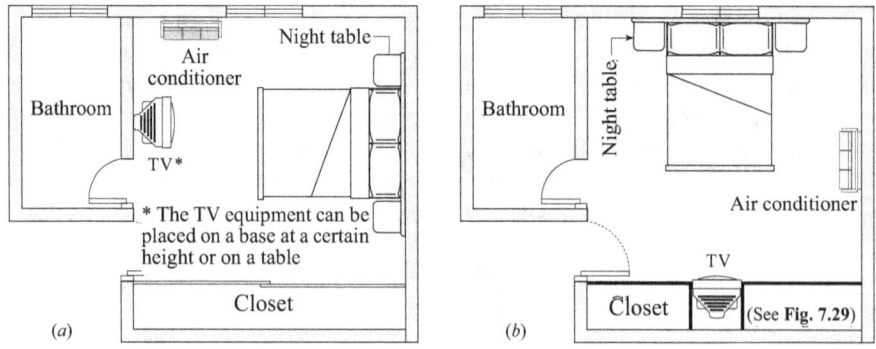

Fig. 118 Spatial distribution in a bedroom with a double bed.

In both cases electrical outlets must be placed on the sides of the double bed to be able to comfortably connect lighting lamps on the night tables.

If instead of a double bed two single beds are used, **Fig. 119** shows two possibilities of location. It is noted in this case that a single outlet could be used to power the two Night tale. The outlets for the air conditioner and for the television remain in the same location as in **Fig. 118**.

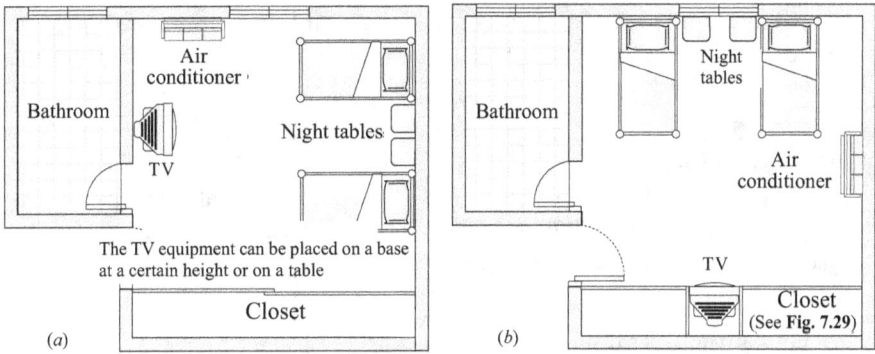

Fig. 119 Spatial distribution in a bedroom with two single beds.

Based on the spatial distributions shown in the previous figures, we will proceed to place the electrical outlets in the bedroom in **Fig. 118**. This gives rise to **Fig. 120**, where the two alternatives already described have been considered. To start placing the electrical outlets there are two possibilities: a) Start from the entrance door and, from there, locate them following the rules studied, adhering to the rule that the distance between them is not greater 3.6 m between two outlets. b) The electrical outlets are located taking into account the probable distribution of the furniture and equipment, and, subsequently, the established regulations are applied in order to make the necessary corrections when the requirements therein are not met. The second option will be adopted as design criteria and we will analyze the two cases below.

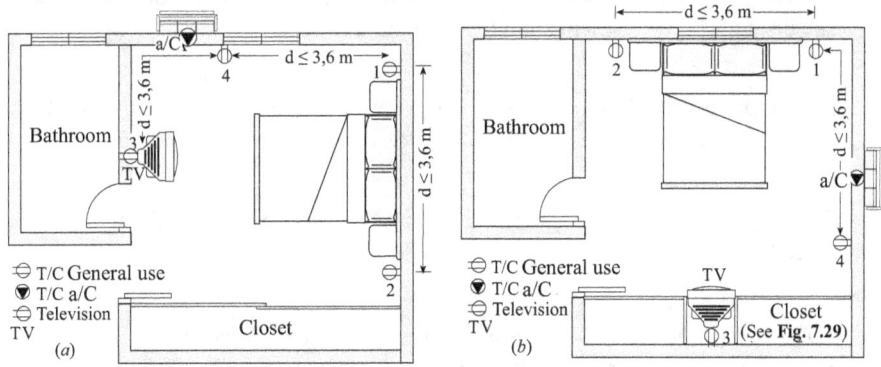

Fig. 120 Distribution of electrical outlets in the bedroom of the **Fig. 118**.

For **Fig. 120**(a), the following steps will be followed:

1. *Tomacorrientes de uso general*: We begin by locating outlets 1 and 2, for general use, on both sides of the double bed, taking into account that the distance

between them should not be greater than 3.6 m. Outlets are placed at a height of 30 cm above the finished floor. These outlets can be used both for bedside lamps and for connecting any appliance in the bedroom: an electric shaver, a vacuum cleaner, etc. General purpose outlets could be split phase, mentioned in the previous section on outlets and switches, which allow the same outlet to be fed with two different phases. This would ensure that, when a failure occurs in one of the phases, the service is maintained. This is not required and is left to the discretion of the designer.

Then, we locate outlet 3 for the television which is placed in front of the bed and at a height that will depend on the owner's decision. If an aerial base is placed the height should be around 1.70 m, while if a table is used as a TV stand the outlet can be left 30 cm above the finished floor.

Outlet 4 is placed at a distance less than or equal to 3.6 m from outlet 1. The spaces corresponding to the entrance door and the closet are not taken into account for the purposes of measuring these distances. According to the latter, outlets 1 and 4 should have a distance of less than 3.6 m from each other.

2. *Individual puylet*: aC symbolizes the outlet for air conditioning. Its height above the finished floor is approximately 2 m. If it is a window air conditioner, this outlet can be 120V, 220, or 240V. If it is a console (evaporator), the outlet voltage is typically 208 or 240V.

For **Fig. 120**(b) the steps are these:

1. *General use outlets*: We start by locating outlets 1 and 2 on both sides of the bed. The distance between them should not be more than 3.6 m.

The TV outlet 3 is placed, built into the closet, at a convenient height in front of the bed. The cabinet in the closet must be designed to house the television.

Outlets 1, 2 and 4 will be distributed so that the distance between them does not exceed 3.6 m.

2. *Individual outlet*: The aC outlet of the air conditioner is located at a convenient height and with a voltage suitable for the selected equipment.

In the case of **Fig. 119**, the bedroom is occupied by two single beds, giving rise to the electrical distributions of **Fig. 121** (see next page). For **Fig. 121**(a), we follow the steps below:

1. *General use outlet*: Outlet 1 is placed between the two night table, ensuring that single beds are served by that outlet.

From socket 1, outputs 2 and 3 are placed at distances d ≤ 3.6 m.

TV socket 4 is placed in front of the two beds and its position above the floor follows the above.

2. *Individual outlet*: The outlet for the air conditioner (aC socket) is placed as described before.

The distribution of outlets in **Fig. 121**(b) follows the same criteria outlined for the previous cases. For these last two cases, split phase receptacles are used.

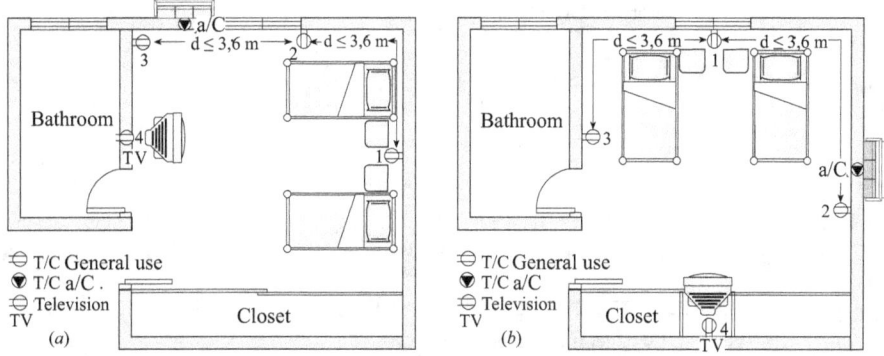

Fig. 121 Distribution of electrical outlets in the bedroom in **Fig. 119**.

It should be emphasized, once again, the advisability of talking with the architect and the owner of the residence to decide, definitively, on the final location of the electrical outlets.

Fig. 122 summarizes the four alternatives presented above and that establish the location of the electrical outlets in a room in the case of double or single beds. In the design of the electrical installation of the bedrooms, the provisions regarding the protection of the branch circuits that feed the rooms must be remembered. There it is established that, for safety reasons all these branch circuits must be protected with arc fault interrupters (AFCI).

Let's pay attention to the lighting in the bedrooms for which we will continue using the room shown in the previous figures. In relation to it, we point out the following aspects:

1. At least one output controlled by its switch should be installed for lighting in the bedrooms. For convenience the bedroom lamp which can be on the ceiling or on the wall*, depending on what has been discussed with the architect and the owner of the work, can be controlled by means of three-way switches from the entrance,

---

* If the ceiling is square, generally the luminaire is installed in the center of the room; if it is made of wood, the luminaires are generally wall sconces.

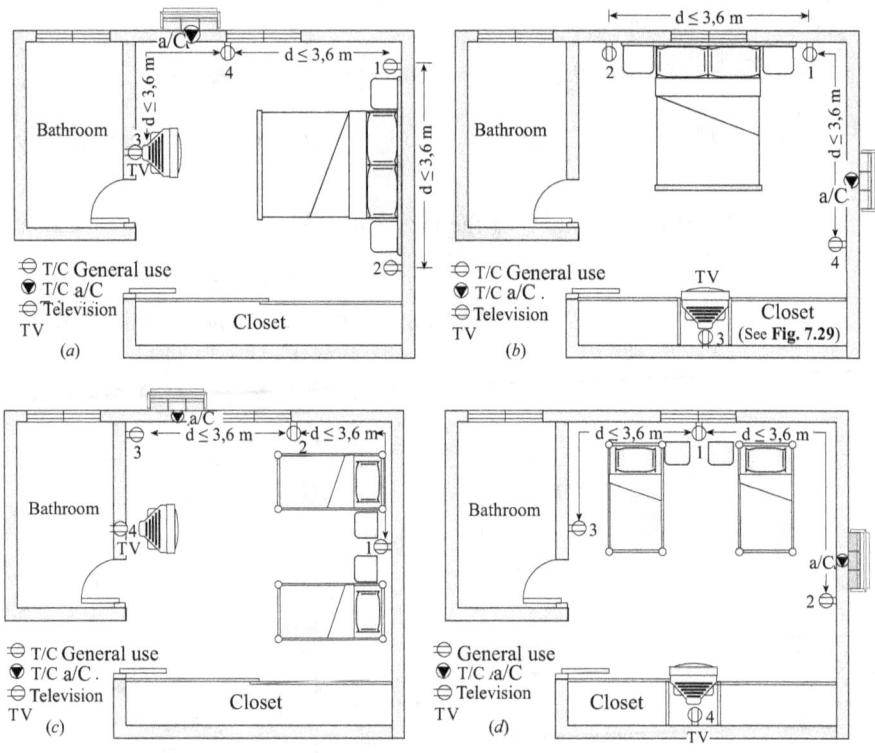

T/C General use
T/C a/C
Television
TV
(a)

T/C General use
T/C a/C.
Television
TV
(b)
Closet (See **Fig. 7.29**)

T/C General use
T/C a/C.
Television
TV
(c)
Closet

General use
T/C /a/C
Television
TV
(d)
Closet

Note: The TV outlet is for general use.

Fig. 122 Summary of alternatives for the location of electrical outlets in a bedroom with double or single beds.

from the room and from a point close to the double bed or between the two single beds. Thus, once the person has gone to bed, can turn off the lamp from his place of rest. Likewise, when someone enters he can turn on the lamp in the room.

2. To improve the security of the house, inside the main room there must be a switch that controls the external lighting of the residence. When the presence of strangers in the residential premises is suspected, it would be possible, turning on the external lights, to dissuade them from intruding the residence.

3. It could be important that the closets inside the room are illuminated to facilitate the selection of clothes by the user. This requires complying with the requirements established in this regard, which start from the definition of what is an area or storage space of a closet, with the purpose of locating the luminaires inside it. The storage space contains clothes, shoes and other clothing items, and is determined as illustrated in **Fig. 123**. At the bottom of the closet the storage space is 60 cm wide from the inside wall of the closet. and 1.80 m high at the height of the tube where the clothes are hung. In the cabinets above the storage space is a minimum of 30 cm wide or the actual width of the cabinet if it exceeds 30 cm.

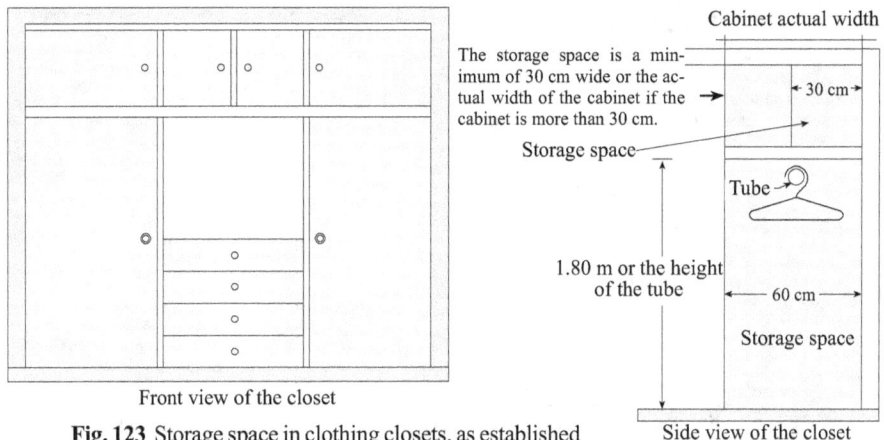

Cabinet actual width

The storage space is a minimum of 30 cm wide or the actual width of the cabinet if the cabinet is more than 30 cm.

Storage space

1.80 m or the height of the tube

Tube

30 cm

60 cm

Storage space

Front view of the closet

Side view of the closet

Fig. 123 Storage space in clothing closets, as established by electrical codes.

The installation of incandescent light fixtures with partially or fully exposed bulbs, attached to or hanging from the ceiling or wall surfaces, is not allowed in the closet (see **Fig. 124**).

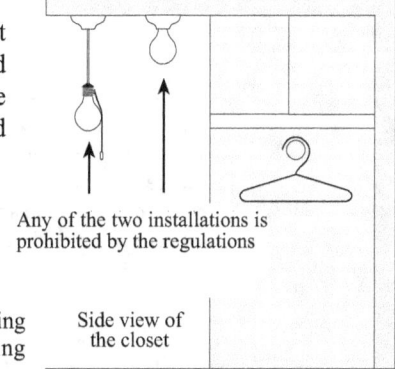

Any of the two installations is prohibited by the regulations

Fig. 124 Violations of the rules regarding the placement of exposed or hanging incandescent luminaires.

Side view of the closet

The installation of luminaires in the closets is allowed according to the following criteria:

• Incandescent luminaires, surface mounted on the ceiling or on the wall, as long as there is a minimum space of 30 cm between the luminaire and the storage space (see **Fig. 125**).

• Fluorescent luminaires, surface mounted on the ceiling or on the wall, as long as a minimum space of 15 cm is left between the luminaire and the storage space (see **Fig. 125**).

• Incandescent luminaires recessed into the ceiling or wall, with a fully enclosed lamp, provided there is a minimum space of 15 cm between the luminaire and the storage space (see **Fig. 126**).

• Fluorescent fixtures recessed into the ceiling or wall, with a fully enclosed lamp, provided there is a minimum space of 15 cm between the fixture and the storage space (see **Fig. 126**).

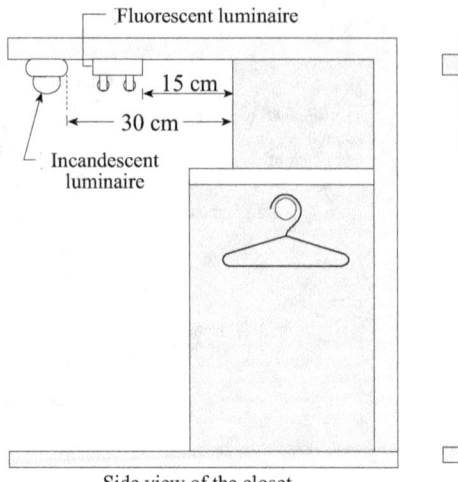

Side view of the closet

Fig. 125 Minimum distances to locate superficial, incandescent or fluorescent luminaires in a clothing closet.

Fig. 126 Minimum distances to locate recessed, incandescent or fluorescent lighting fixtures in a clothing closet.

It must be reiterated that in the final design of an electrical installation it is very important to exchange opinions with the architect of the building and its owners. In this way, an optimal engineering project will be achieved.

In the plans corresponding to the electrical design a variety of symbols are frequently used, indicative of the different elements that will make up the electrical installation. These symbols include the following:

S   Simple switch

S_3 Three way switch

⊖   Output for luminaire

Fan + luminaire

———   Conduit in ceiling or wall

- - - - - -   Recessed conduit in floor

As we progress through the design of residential electrical installations we will add other electrical symbols.

Let's take the furniture distribution model shown in **Fig. 127**, in which the double bed or the single beds appear placed in front of the wall where the window is located.

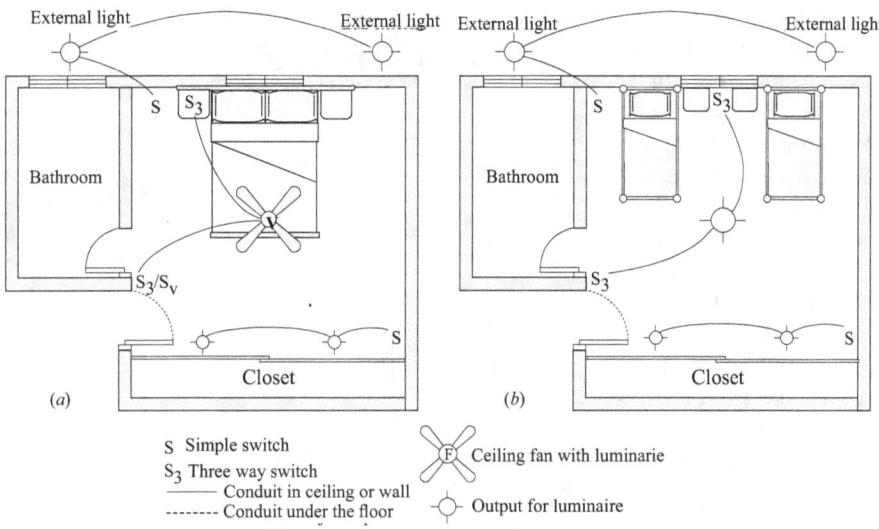

Fig. 127 Lighting and control of luminaires in a bedroom.

Let's see the details of the arrangement of luminaires and their switches in the bedroom:

Figura 127(*a*): A fan with light is placed in the center of the room with a control (switch S) for the fan at the entrance to the room (the single switch could also be placed next to the bed). The luminaire incorporated into the fan is controlled from the entrance door and from the bed through the two S₃ three-way switches.

Las dos luces externas se controlan desde el interior de la habitación mediante el interruptor sencillo S.

The luminaires, placed in front of the closet doors and following the rules already studied regarding their location, are controlled with a simple switch S placed on the wall of the room. It is also possible to use a switch that activates when the closet door is opened.

Figura 127(*b*): Instead of a ceiling fan with a built-in light fixture, a ceiling light (fluorescent or led) is used. The other lamps and their switches are distributed in a similar way to that of the **Fig. 127**(*a*).

26. ELECTRICAL OUTPUTS IN THE ROOM

Fig. 128 shows the location of outlets and outputs for the luminaires. Four general purpose outlets are proposed in the room, one of which can be used for the TV. Also, an output is left for the connection of an air conditioner or a Split-type equipment console.

The room is illuminated by two luminaires with ceiling outputs and controlled by three-way switches placed at the entrance door to the house and at the entrance to the room from the rooms. This is very convenient for access to the room, whose lights can be turned on from two different points. The garage luminaires are controlled by a switch S from inside the room. The luminaires of the porch are controlled with the S switch from the same porch. This allows that a person arriving at the entrance door can easily see the lock. Porch and garage lights could also be controlled by three-way switches from inside and outside the residence. Note that new symbols have been added in the electrical system diagram.

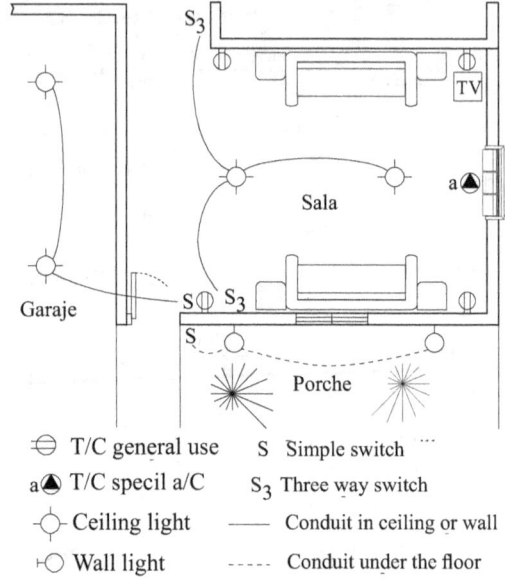

⊖ T/C general use   S Simple switch
a▲ T/C specil a/C   S₃ Three way switch
-○- Ceiling light   —— Conduit in ceiling or wall
⊢○ Wall light   ----- Conduit under the floor

Note: One of the T/C of the room will be used for the TV.

Fig. 128 Distribution of luminaires ∖T/C general use˙ and outlet switches in the living room.

27. ELECTRICAL OUTLETS IN THE LAUNDRY ROOM

It is common to locate the laundry room in a space where all the cleaning and ironing tasks are carried out. Among the typical electrical appliances in this environment, which use individual outlets, we find:

• Washing machine   • Dryer   • Iron   • Water heater

A layout for the laundry area is shown in **Fig. 129**. A luminaire is used in the center of the area, controlled by a switch at the entrance. Individual electrical outlets are installed to connect the iron, the water heater, the washer and the dryer. Two utility outlets are used for whatever fixture you want to plug into the laundry room, one behind the door and one halfway up the back wall. Its location minimizes the possibility that, if furniture is placed in this environment, it will be left behind. The following provisions refer to the laundry room:

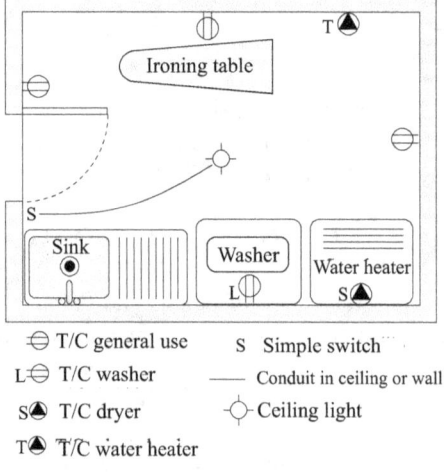

⊖ T/C general use   S Simple switch
L⊖ T/C washer   —— Conduit in ceiling or wall
S▲ T/C dryer   -○- Ceiling light
T▲ T/C water heater

Fig. 129 Lights and outlets in the laundry room.

1. At least one 20A circuit must supply power to the utility outlets. It must not be connected to any other outlet outside the laundry room.

2. There should be at least one electrical outlet in the laundry area. Exceptions are those multi-family dwellings where there are laundry facilities in the same building.

3. There should be at least one electrical outlet in the laundry area. Exceptions are those multi-family dwellings where there are laundry facilities in the same building.

4. The utility outlets, located at a distance of less than 1.8 m from the external side of the tub, must be of the GFCI type (see **Fig. 130**).

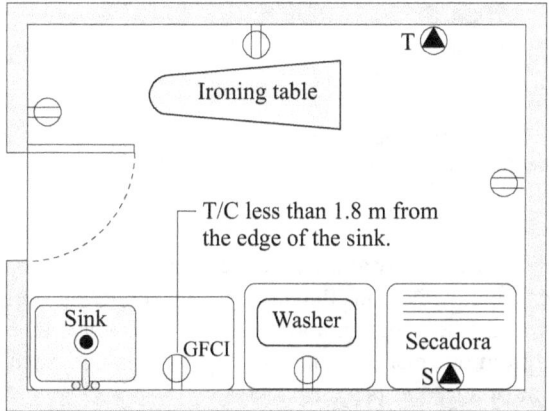

Fig. 130 Electrical outlets located less than 1.8 m from the edge of the sink must be of the GFCI type.

In some cases the washer and dryer are combined in order to save space in the laundry room. In this case, an individual outlet must be reserved for the set.

28. ELECTRICAL OUTLETS IN CORRIDORS

Every corridor of length greater than or equal to 3 m must have at least one electrical outlet. The length of the aisle refers to the length of the center line of the aisle. In addition, at least one outlet for light must be installed, controlled by a switch. See **Fig. 131**.

29. ELECTRICAL OUTLETS IN THE GARAGE

At least one outlet must be installed in the single-family home garage. All 15 and 20 amp single phase receptacles installed in the garage are required to be of the GFCI type for personnel protection (see **Fig. 132**).

Excepted from the above requirement there are power outlets that are not easily accessible, such as those used for controls that automatically open vertical sliding garage

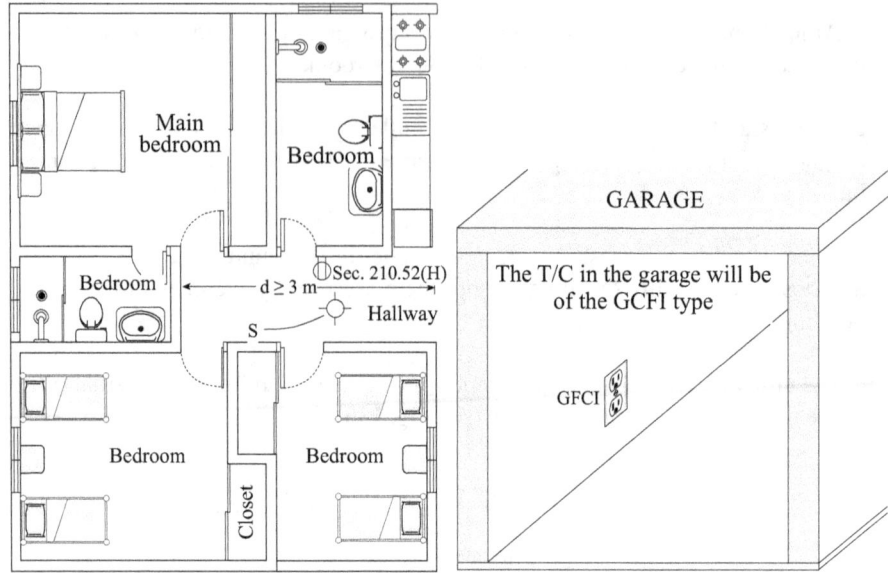

Fig. 131 In the corridors at least one electrical outlet and one outlet for a lamp, controlled by a switch, must be left.

Fig. 132 The electrical outlets in the garage, with the exceptions cited in the text, must be of the GFCI type.

doors. These outlets are typically attached to the garage ceiling near the engine and its controls*. Also excepted are outlets intended to connect permanently and in a prede-termined space to equipment that, in normal use and connected by plug and cord, are not easily moved from one place to another within the garage, such as a refrigerator, a washing machine or a freezer (see **Fig. 133**).

General purpose (non-GFCI) outlets may be used when electrical appliances are placed in a space dedicated to them; they connect via plug and cord inside the garage and cannot be easily moved. The T/C can be single or double.

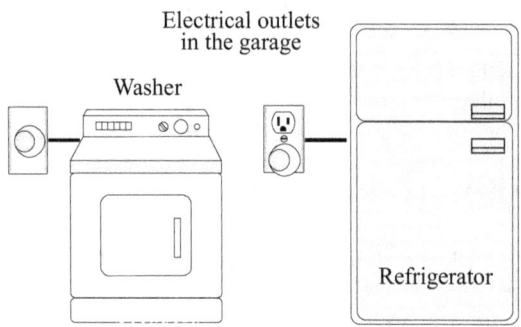

Fig. 133 Use of specific outlets in a garage.

* The **National Electrical Code** states that all outlets in the garage will be of the GFCI type.

As for lighting, you have to pay attention to the placement of the car inside the garage. If it is a garage for one vehicle, two lighting outlets must be left on both sides of it, starting from the central axis and at the end of the garage. This ensures good lighting near the hood in case any service is required to the car. The lamps to be placed can be fluorescent. In the same way, it is convenient to place, in the central area of the garage, a lamp outlet that allows lighting it in normal cases, when it is not necessary to concentrate the light on the hood.

Fig. 134 indicates the placement of electrical outlets and lighting outlets in a one-vehicle garage. Two electrical outlets were placed at the beginning of the garage and two at the end of it. The latter allow you to connect any electrical device that is going to be used to repair the car. Outlet 1 is a GFCI type and must be cascaded to outlets 2, 3, and 4. Individual appliances such as refrigerators or washing machines are not intended to be used in the garage and therefore no outlets are provided for such use. Although not shown in this figure, it is necessary to foresee the installation of a system to automatically close the garage door. This system can close the door by sliding vertically, in which case the outlet can be placed on the ceiling and does not need to be of the GFCI type, or it can also be horizontal and, if the outlet is inside the garage area, it must be used a GFCI. Notice that when the layout of the outlets in a room was described (**Fig. 128**), a simple switch was placed to control the lights in the garage. The new design presented now perfects the previous one.

Two outlets for lamps were placed at the end of the garage to facilitate the visualization of the vehicle's engine. These outputs are controlled by three-way switches located inside the house and at the entrance to the garage. Likewise, an outlet for a lamp (L_3)

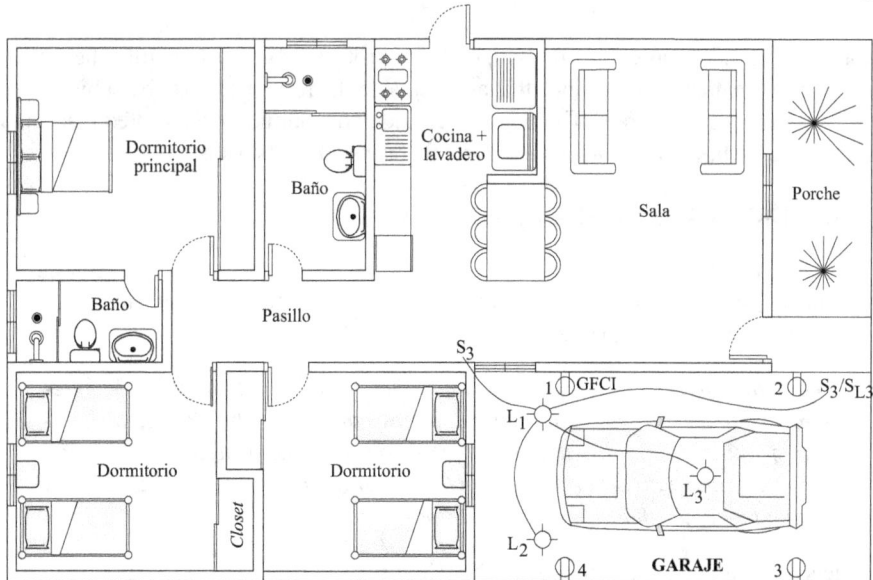

Los T/C 2, 3 y 4 se conectan en cascada al T/C 1, que es del tipo GFCI, por lo que actúa como protección de los primeros.

Fig. 134 Tomacorrientes y salidas de iluminación en un garaje.

was left, located near the middle of the garage and controlled by a simple switch, SL_3, installed in the same box as S_3. This light provides general lighting in case it is not required to turn on the lamps L_1 and L_2.

30. ELECTRICAL OUTLETS ON THE PORCH

On the porch there are (see **Fig. 135**) two GFCI type outlets because they are outlets outside the house. One of these outlets is controlled by a switch located in the living room and could be used to connect a Christmas tree or decorative lights at Christmas.

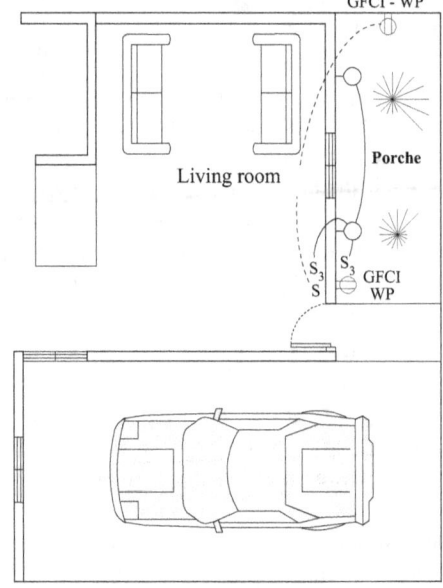

The porch light fixtures are controlled by two three-way switches. This has a double purpose. A person who arrives at the residence at night can turn on the lamp in order to see the door lock. On the other hand, whoever is inside the house can turn on the porch light to observe what is happening outside the residence without having to open the front door.

Fig. 135 Electrical outlets on the porch.

Note that the electrical outlets on the porch are protected against humidity (hence the letters WP: weather proof), since this area, outside the residence, may be subjected to climatic factors such as rain. The porch is considered a humid, open, roofed place and not exposed to the beating rain or the water that runs down its walls.

31. EXTERNAL EXITS TO A RESIDENCE

In all single-family dwellings that are at ground level, at least two electrical outlets shall be installed outside a dwelling, one at the front and one at the rear, at a height not greater than 2 m above the ground.

The above rule states that the minimum number of outlets outside a single-family residence is two. That is, more than two outlets can be placed, the location of which will be distributed at the discretion of the installation designer, in consultation with the architect and the owner of the residence. These outlets cannot be higher than 2 m above the floor.

Outlets for outdoor outles must be of the GFCI type.

This last statement should not be put aside in a residential electrical installation, since failure to observe it can lead to fatal accidents. Even those outlets that are higher than 2 m (for example, those used for Christmas lights or on balconies of a residence), but outside the house, must be of the GFCI type.

It is important to mention again the type of outlet and its accessories when we are in the presence of humid (damp) and wet (wet) places, spaces that are generally found outside a residence:

Damp Locations: An outlet installed outside of a residence, protected from the weather or in another damp location, must have a weatherproof cover when the outlet is covered (with no plug inserted and cover closed)..

An outlet shall be considered weatherproof when it is located indoors in open porches, domes, and the like, and is not exposed to driving rain or running water on the surfaces that house it.

In accordance with the foregoing, electrical outlets located indoors, on the exterior of a residence, must have an external cover and weatherproof cover.

Wet Locations: 15 and 20 amp and 125 volt and 250 volt circuit outlets must have weatherproof covers whether or not the equipment plug is connected.

All other outlets installed in wet locations must comply with the following:

(*a*) An outlet installed in a wet location, where the appliance to be connected is not supervised while in use, shall have a weatherproof cover with the plug inserted or not.

(*b*) An outlet installed in a wet location, where the appliance to be plugged in is watched while in use (for example, portable tools), shall have a weatherproof cover when the plug is not plugged in.

The outlets listed above as unattended include, but are not limited to, those used to power Christmas lights, water pumps, and some of the garage door openers.

The design of the electrical installation in the exteriors of a residence requires a detailed study of the possible devices to be connected. The existence of different environments and possibilities outside the interior of a home creates various design alternatives. Hence the need to explore all the possibilities of comfort, in close collaboration with the person in charge of the architectural project and the owner. The most sophisticated homes consist, among others, of spaces such as swimming pools, lighting poles, barbecue grills, huts and reflectors, which need the appropriate electrical outlets. Also, the electrical outlets of some air conditioners are placed outside of houses.

From the foregoing it can be deduced that external electrical outlets, subject to the inclement of environmental factors, not only have to be of the GFCI type, but also need to be protected against the penetration of water inside. A typical assembly for an outlet placed outside a home is shown in **Fig. 136**, as well as the depths to which outlets should be buried.

It can be seen in the **Fig. 136** that two tubes are used to support the socket: one of them houses the UF cable internally, while the other is empty and only gives stability to the structure. It is recalled that the UF cable is resistant to moisture and heat and can be buried directly in the ground. It is flame retardant and can be exposed directly to the sun.

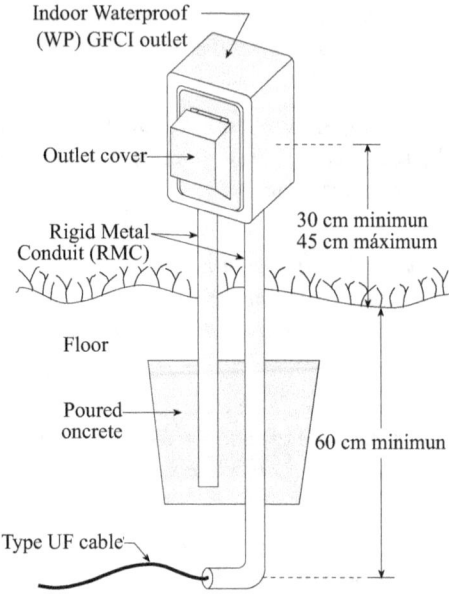

Fig. 136 Structure for mounting an electrical outlet on the ground outside a home.

It can be seen in the previous figure that two tubes are used to support the socket: one of them houses the UF cable internally, while the other is empty and only gives stability to the structure. It is recalled that the UF cable is resistant to moisture and heat and can be buried directly in the ground. It is flame retardant and can be exposed directly to the sun.

Let's look at some aspects of external lighting. Lighting, apart from providing security during night hours, becomes an important decoration factor in patios and gardens. We already discussed the location of the luminaires in the porch, which is also an external area. Other external areas have to be studied individually to decide the outputs for the luminaires.

External lighting, especially in gardens exposed to the elements, at the mercy of water and other deteriorating agents, requires the use of appropriate luminaires for such environments, which must be manufactured and authorized for places that, under rainy, snowy conditions or irrigation, do not allow the penetration of water inside. Likewise, the switches to control the external lights, if they are placed outdoors, must be of the type resistant to the same (WP).

For general purpose exterior lighting, reflectors (R) or Parabolic Aluminized Reflectors (PAR) are suitable lamps that come in sets of one to three. In particular, the latter are not affected by climatic factors. The sockets of these lamps must be waterproof. It is also common to use conveniently placed lighting poles.

In the case of decorative lights, 120 V lamps and 12 V low voltage units are on the market. If you decide to install a low voltage installation, you will need a 120 V to 12 V step-down transformer, which is placed on the external wall of the dwelling (see **Fig. 137**).

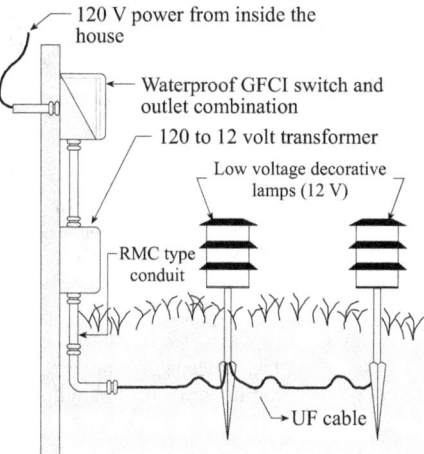

Fig. 137 Low voltage outdoor lighting.

Table 5 specifies the minimum depths to which cables or conduits used in patios and gardens must be buried as far as a residence is concerned.

Installation type	Column 1 Conduit RMC or IMC	Column 2 Conductors or cables directly buried	Column 3 Non-metallic conduits to be directly buried without being embedded in concrete.	Column 4 Residential branch circuits rated 120V or less with GFCI protection and 20A maximum overcurrent protection.	Column 5 Irrigation and circuits limited to no more than 30 V and installed with UF cable or with another type of cable or conduit.
Conduits not placed in trenches.	60 cm	15 cm	45 cm	30 cm	30 cm
Conduits in trenches under a 50 mm thick layer of concrete.	45 cm	15 cm	30 cm	5 cm	5 cm

Tabla 5 Minimum depths to which cables or conduits used in patios and gardens must be buried.

The use of trees as means of support for external luminaires is permissible. However, the use of trees as supports for pendant conductors is prohibited. The two situations are illustrated in **Fig. 138**.

Based on what has been discussed so far in this section we conclude that the design of electrical installations outside a home requires careful planning, which includes factors

Reflectors

Conduit

Allowed: xternal decorative lights that use a tree as a su-pport.

Violation: anging cables are not allowed to use the trees for support.

Fig. 138 Trees can be used as a means of supporting light fixtures, but are not not allowed for support of hanging cables.

related to personal safety and electrical safety, with the need to locate outlets in places appropriate and with the beautification of gardens, walkways and recreational spaces within the limits of the property.

The regulations used in this book follow the US and electrical codes of many Latin American countries. The concepts studied are, in general, applicable to most nations. Although there may be some variations with respect to what is discussed in this book, according to specific regulations of different regions, much of what is studied can be adapted to other electrical regulations.

As an example of the distribution of electrical outlets outside a residence we present **Fig. 139** (see next page). It is an isolated single-family home with three bedrooms and two bathrooms, kitchen and living-dining room. Externally it has a covered porch on the front with two small garden areas and a backyard, where the compressors that supply air conditioning to the living room and bedrooms are located. The house is fenced with block walls and the side areas are open, and each of them can be used as a garage or as a place of recreation. Let's analyze the front, side and rear parts of the outdoor electrical installation.

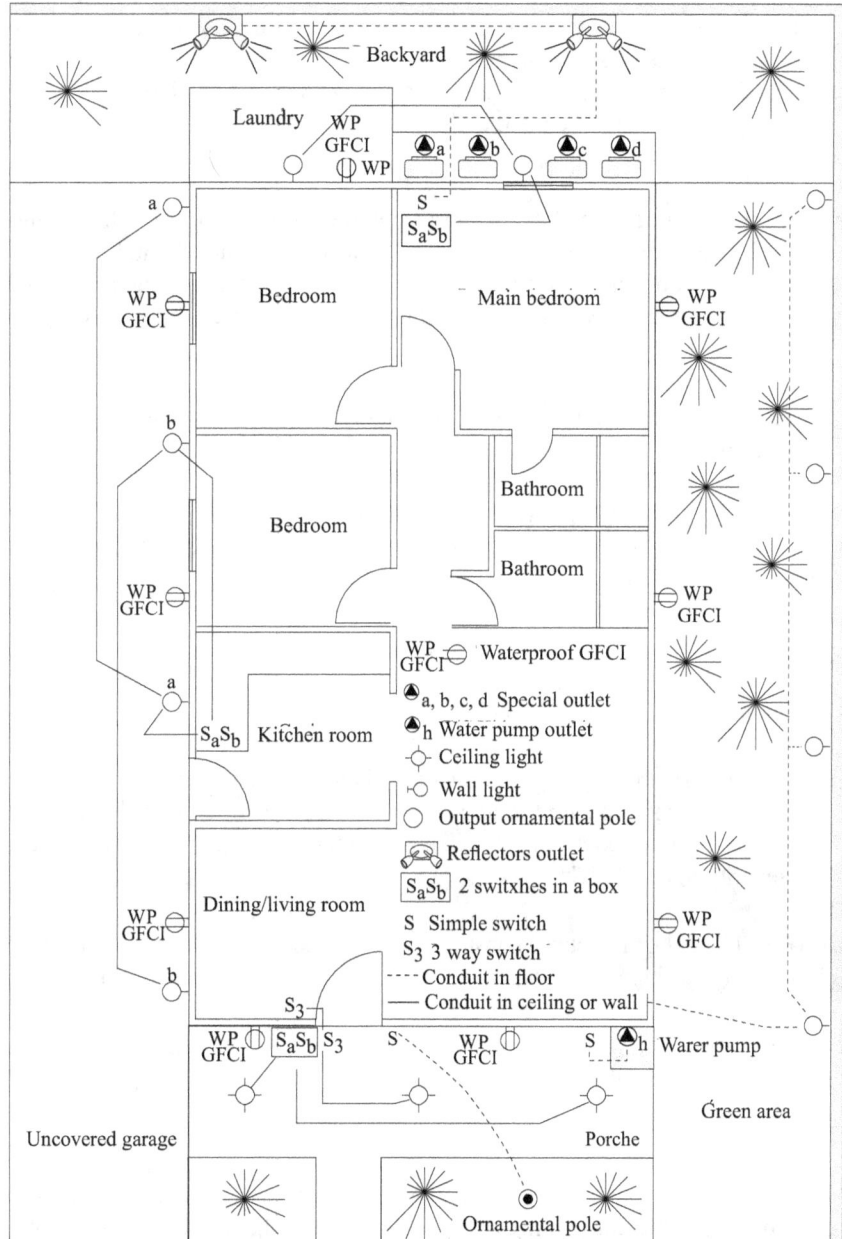

Fig. 139 Example of electrical outlet layout outside a residence

Front part (porch)

We have the following items:

a) An ornamental post, placed in the garden and controlled by switch S that is
located next to the front door.

b) A lamp located in the center of the porch controlled by two three-way switches which can turn it on or off externally, from the porch, or internally, from the living room. The idea behind this double check is that a person arriving at night can easily insert the key into the lock. Also, if you want to light the porch without leaving the house use the internal three-way switch.

c) The two lamps at the ends of the porch complete the trio of lights that illuminates it. They are individually controlled by a double switch, S_a and S_b, placed next to the door. In this way one to three lamps can be turned guaranteeing good lighting and savings in consumption when complete light coverage is not required.

d) Two general purpose, waterproof and ground fault protected outlets are located on the porch.

e) Since the hydropneumatic system (water pump) is located on the porch a special outlet with a switch that controls the motor of the equipment is placed on it. In this case it is a 120V 1/2 HP single phase motor.

Backyard

The following elements are distinguished:

a) We have two wall lamps that illuminate the laundry area and the cemented area, where the compressors for the air conditioners in the bedrooms and living room are. These outlets, controlled by a double switch in the main room, also serve as security lights at night.

b) A GFCI breaker, protected against moisture, is placed in the laundry area where a washing machine or any other electrical appliance can be connected.

c) Two double floodlights controlled by a simple switch from the main room are used to illuminate the backyard.

Left side (garage)

The left side surface can be used as a garage or as a recreation area; Hence, it is necessary to place sufficient lighting and electrical outlets that allow the connection of sound equipment and electrical devices:

a) There are four external wall luminaires, controlled, in groups of two, by a double switch placed in the kitchen. This saves energy by having the option of not turning on all the lamps at the same time.

b) Three waterproof GFCI outlets are located throughout the garage.

Right side (green area)

The right side surface can be used as a garage or as a green area. The electrical elements that stand out are:

a) Four wall lamps illuminate the area along it. These lights are controlled from the room. They can all be turned off at once or a double switch can be used to control them in pairs.

b) Three waterproof GFCI outlets are positioned to supply power to whatever electrical fixture you wish to plug into that space.

32. ELECTRICAL OUTLETS ON STAIRS

On the two levels that are linked by a stair-case, switches must be placed to control the lighting of the same. An important role here is played by three-way switches, which will allow the lamp to be switched on or off from the lower and upper levels of the staircase, as shown in **Fig. 140**.

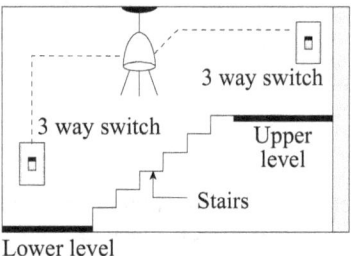

Fig. 140 Stair lighting.

33. EXITS AROUND BODIES OF WATER

The surroundings of bodies of water, such as swimming pools, water fountains and jacuzzis, are places particularly sensitive to electrical accidents, since they combine electric current with humidity and the absence of footwear as part of the usual cloth-ing of the human being. Severe limitations have been placed on electrical outlets and lighting around swimming pools, water fountains, and other similar facilities. No general purpose outlet shall be within 6 feet of the inside wall of a swimming pool, water feature, or hot tub. This prevents any electrical appliance from being connected to the outlet and at a very close distance to water, and thus reduces the risks of electrical accidents. **See Fig. 141**.

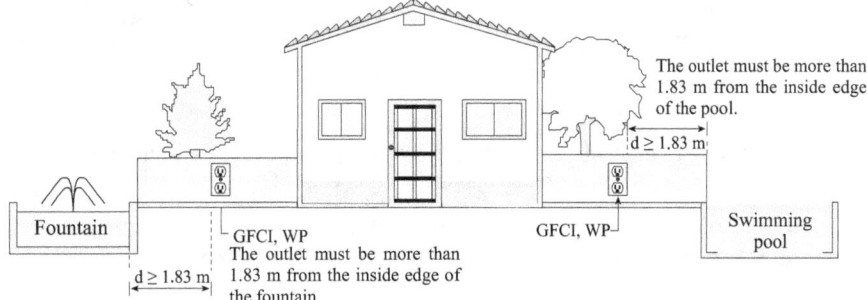

Fig. 141 No general purpose outlet should be within 1,83 m of the inside wall of a pool or water feature.

Likewise, the regulations require that at least one 15 or 20 A and 125 V outlet be placed in a dwelling at a distance between 1.83 m and 6 m from the internal wall of a permanently installed swimming pool, as shown in **Fig. 142**. These outlets must not have a height higher than 2 m above the floor level.

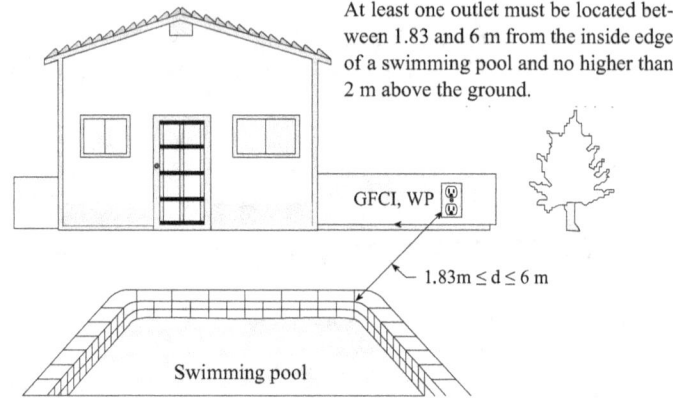

At least one outlet must be located between 1.83 and 6 m from the inside edge of a swimming pool and no higher than 2 m above the ground.

GFCI, WP

1.83m ≤ d ≤ 6 m

Swimming pool

Fig. 142 There must be an electrical outlet close to the pool, at a distance that is between 1.83 m and 6 m from the internal edge of the same.

In relation to the water pumps used in swimming pools it is established that the outlet to which a pump is connected will be at a minimum distance of 3 m from the internal wall of the pool. It is allowed to place this outlet at a distance of not less than 1.83 m if it meets the following conditions: (1) The outlet is simple. (2) It is of of lock type. (3) Can be grounded. (4) Has GFCI protection. See **Fig. 143**.

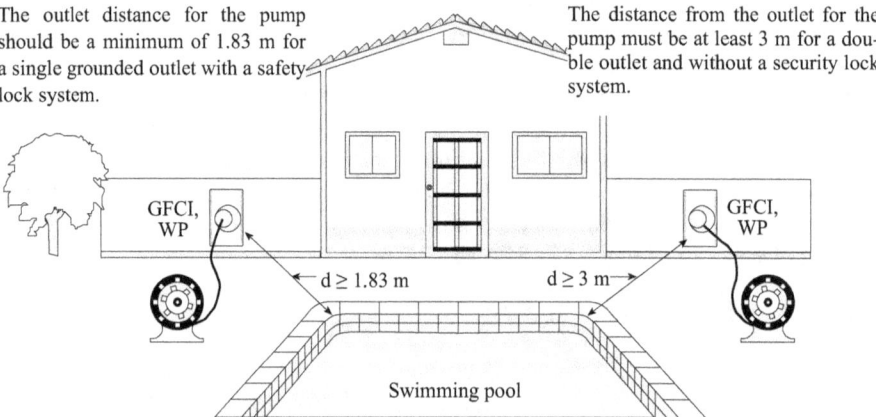

The outlet distance for the pump should be a minimum of 1.83 m for a single grounded outlet with a safety lock system.

The distance from the outlet for the pump must be at least 3 m for a double outlet and without a security lock system.

GFCI, WP

GFCI, WP

d ≥ 1.83 m

d ≥ 3 m

Swimming pool

Fig. 143 Limitations of the distance between the outlet for the connection of the water pump and the internal edge of the pool.

It is allowed to illuminate swimming pools as long as certain requirements are met. The most relevant in terms of swimming pools are the following:

1. *Luminaires above swimming pools located outside a residence shall be installed outside an area extending horizontally 1.5 m from the internal walls of the swimming*

pool and at a height of not less than 3.7 m above the water level. See **Fig. 144**.

2. *Switches near a swimming pool shall be located at least 1.5 m measured horizontally from its internal wall unless separated by a solid fence, wall or other permanent barrier. The installation of a switch is allowed at a distance less than that mentioned as long as it appears to be suitable for this use. See* **Fig. 145**.

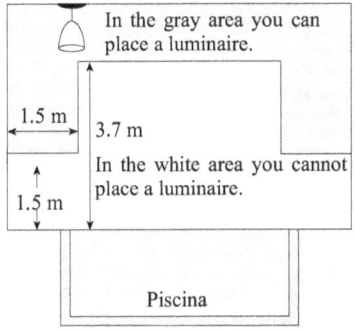

Fig. 144 Delimitation of the area where a luminaire can be placed on top of a swimming pool.

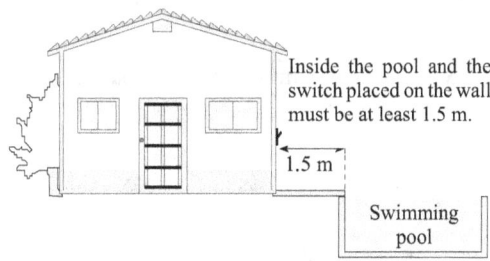

Fig. 145 Delimitation of the distance between a switch and the internal edge of a swimming pool.

Think...
Explain...

---

70. Explain the convenience of discussing with the owner of a home the electrical project of the same.

71. Is there a discrimination between homes for people of different social classes in the regulations that govern the design of electrical projects? Explain why the rules of this regulation are important when designing an electrical project.

72. List the characteristics of a proper and safe electrical design.

73. Mention the steps that must be followed to optimize the design of an electrical installation.

74. Why is it convenient to try to place two electrical outlets facing each other even when they are located in different environments? (**Fig. 27**)?

75. Explain why it is advantageous to place one of the outlets in a room below the switch that controls the light inside the room.

76. Can a bedroom light switch be placed behind your front door? Is this convenient?

77. How do you define a wall space? Is this definition important?

78 What is the minimum wall space in which an electrical outlet must be installed in a home?

79. Can sliding glass doors be considered a wall gap?

80. What is the minimum distance between outlets in environments such as bedrooms, living room and dining room? How does this relate to the requirement that no point, measured horizontally on the floor line, be more than 1.80 m from an electrical outlet?

81. Can an electrical outlet be installed on top of a heating equipment? Reason your answer.

82. Should those outlets that are an integral part of a piece of equipment be taken into account to determine the number of outlets to be installed in a home?

83. An electrical outlet is located at a height of 2.20 m. Should this circumstance be taken into account to determine the number of them in a dwelling?

84. How far from the wall must a floor socket be located to be taken into consideration when determining the number of floor sockets to be installed in a dwelling?

85. What is a split phase outlet? How does this type of outlet connect to power leads? Make the corresponding drawing.

86. In the kitchen of a house there is a space in a wall of 40 cm. Is it mandatory to place a general purpose outlet in this space?

87. There is a 2 m corridor inside a residence. Should an electrical outlet be placed in that hallway?

88. At what maximum distance from an air conditioning unit must an electrical outlet be installed?

89. Explain about the electrical outlets in the kitchen room.

90. How many small appliance circuits must feed into the kitchen room outlets?

91. Does a kitchen room refrigerator or freezer have to have an individual outlet? If the answer is positive, what is the reason?

92. Can the small appliance circuit be used to power the general purpose outlets located on the kitchen room walls?

93. Say which of the following pieces of equipment can be connected to the small appliance circuits in the kitchen room:

    a) Blender.   b) Automatic dish washer.   c) Electric clock.
    d) Waste disposer.   e) Kitchen ignition systems.

94. What should be the distance between two consecutive electrical outlets placed on top of cabinets in a kitchen room?

95. Do the spaces occupied by the dishwasher, stove and sink have to be taken into account when measuring the distance between electrical outlets on the top of a kitchen cabinet?

96. Which outlets should be GFCI type in a kitchen room?

97. Is a GFCI type outlet required for a refrigerator or freezer located in the kitchen room?

98. According to what was mentioned in this chapter, which electrical equipment consumes the most energy?

99. Which individual circuits are commonly found in the kitchen room?

100. How do you define a peninsula and island in a kitchen room? What are the requirements established regarding the installation of electrical outlets there?

101. Why are electrical outlets not allowed to be placed face up in kitchen cabinets?

102. How high above kitchen cabinet counters should electrical outlets be located?

103. How can electrical outlets be installed on top of a peninsula or island in a kitchen room?

104. Describe the lighting points to consider in a kitchen room and in the dining room of a house.

105. How many electrical outlets can be installed in a bathroom?

106. In a bathroom an electrical outlet is placed behind the entrance door. Does it have to be a GFCI type?

107. How far from the edges of a sink can an electrical outlet be located?

108. How far below the top of a bathroom cabinet can an electrical outlet be located?

109. Can a branch circuit feed the outlets in a bathroom and the room next to it?

110. What do the rules establish regarding the location of electrical outlets within the shower area?

111. Describe how the lighting should be distributed in a bathroom.

112. Explain the limitations on the placement of light fixtures in the bathroom.

113. Comment on the place where the luminaires are going to be placed in the bathroom in relation to the location of the mirror.

114. Under what conditions are switches allowed inside the bathtub or shower area of a bathroom?

115. In a bedroom, the maximum separation between wall outlets is 3.6 m. How does this standard fit in with the location of the furniture in a room which, eventually, may lead to considering distances smaller than the one indicated?

116. What is the use of three-way switches in bedrooms?

117. Describe the most important points to consider in the design of the electrical installation of a bedroom taking into account the location of outlets and lights with their respective control switches.

118. What are arc fault interrupters and why is their use important in the bedrooms of a home?

119. Describe the characteristic of a split phase socket and the importance they have in an electrical installation.

120. Comment on the usefulness of having switches inside the main room of a residence to control its external lights.

121. Describe the lighting characteristics in the bedrooms of a house.

122. In the lighting of the closet of a room, define what is storage space in relation to the installation of lighting fixtures.

123. Is it allowed to install lamps or incandescent bulbs to illuminate the clothing closets? If not, explain what is the reason for this prohibition.

124. Is it allowed to install fluorescent lamps or bulbs to illuminate the clothing closets? In any case, explain your answer.

125. Mention, in general terms, the electrical outlets that should be in the room with respect to lights, sockets and switches.

126. Draw the electrical symbols used in the architectural diagrams in this book. Research other electrical symbols used on blueprints.

127. List the typical devices used in laundry rooms.

128. Can the circuit that feeds the laundry supply power to outlets located in other rooms?

129. At what maximum distance from a specific appliance should an electrical outlet be located in the laundry room?

130. Do all outlets in the laundry room have to be the GFCI type? Explain.

131. Which outlets in the garage must be the GFCI type?

132. State how lighting should be in a garage with respect to the location of the lamps and in relation to a parked car.

133. Describe what the electrical outlet and lighting system should look like on the porch of a residence. What factors must be taken into account?

134. How many electrical outlets must be placed on the exterior of a residence? Where should they be placed?

135. Which outlets must be of the GFCI type on the outside of a residence? Which ones should be waterproof?

136. What do the regulations say about electrical outlets that must be placed in damp or wet locations outside of a residence?

137. Comment on the decorative lighting that is placed outside a residence, citing the characteristics that the lights must have and the switches to use.

138. Can trees be used as support for wires outside a residence?

139. Can trees be used as support media for light fixtures?

140. Explain the lighting system in the stairs of a house.

141. What restrictions have been established for the distance between electrical outlets and the walls of a swimming pool or a water source?

142. What restrictions have been established for the distance between a switch and a swimming pool located outside a residence?

143. What restrictions exist for the luminaires above a swimming pool.

Practicing your knowledge

In the figures corresponding to the problems, it is advisable to enlarge the drawings in order to place the different elements of the electrical installation. Their scale shown is approximate and, although these are views from above, the reader must imagine the presence of other elements, such as a microwave and a garbage disposal.

27. **Figures 146** to **149** correspond to different distributions of spaces in kitchen and dining rooms. Based on what has been studied in this chapter, place the outlet for electrical outlets, lighting fixtures, and switches.

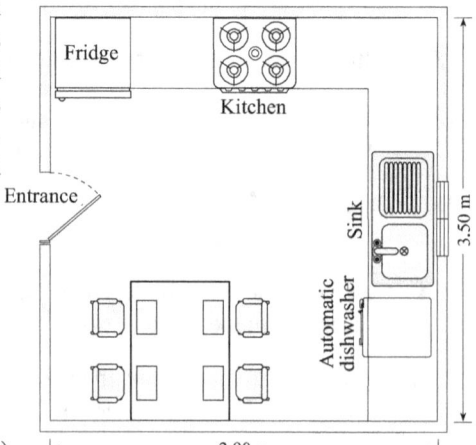

Fig. 146 Ejercicio 27 Part (*a*)

Kitchen

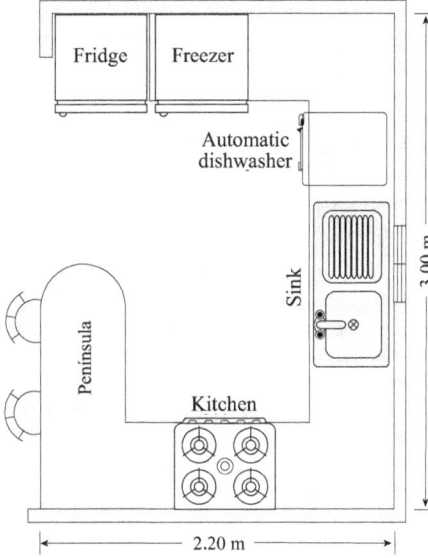

Fig. 147 Exerise 27. Part (*b*)

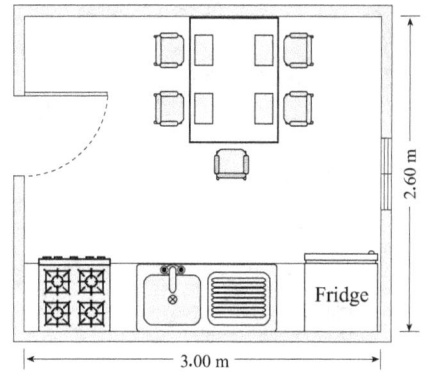

Fig. 148 Exercise 27. Part (*c*)

28. The bathrooms (**Figures 150 to 152** require electrical installations adapted to the provisions of the NEC. Show, on a drawing, the location of electrical and lighting outlets and switches.

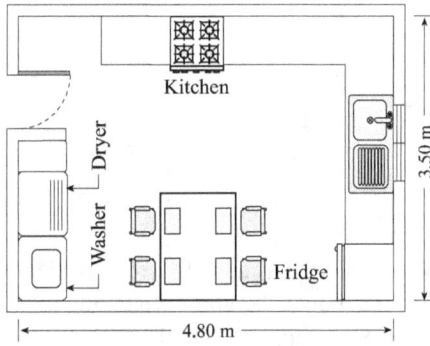

Fig. 149 Ejercise 27. Part (*d*)

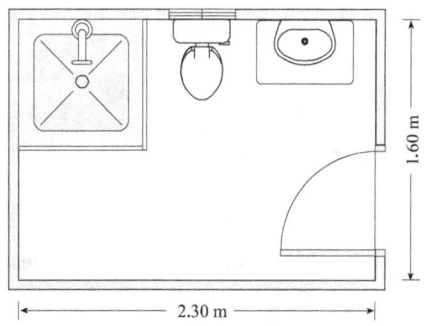

Fig. 150 Exercise 28. Parte (*a*)

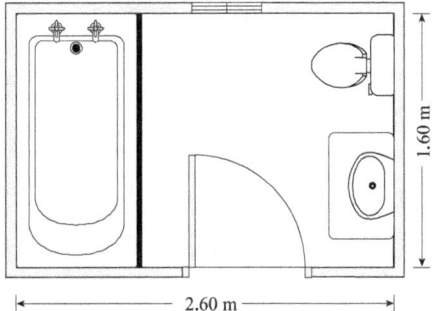

Fig. 151 Excersice 28. Parte (*b*)

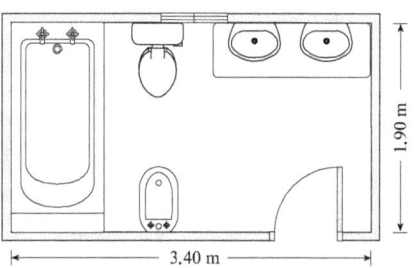

Fig. 152 Exercise 28. Part (*c*)

29. Locate electrical outlets, light fixtures, and switches for the bedrooms in **Figures 153** to **156**.

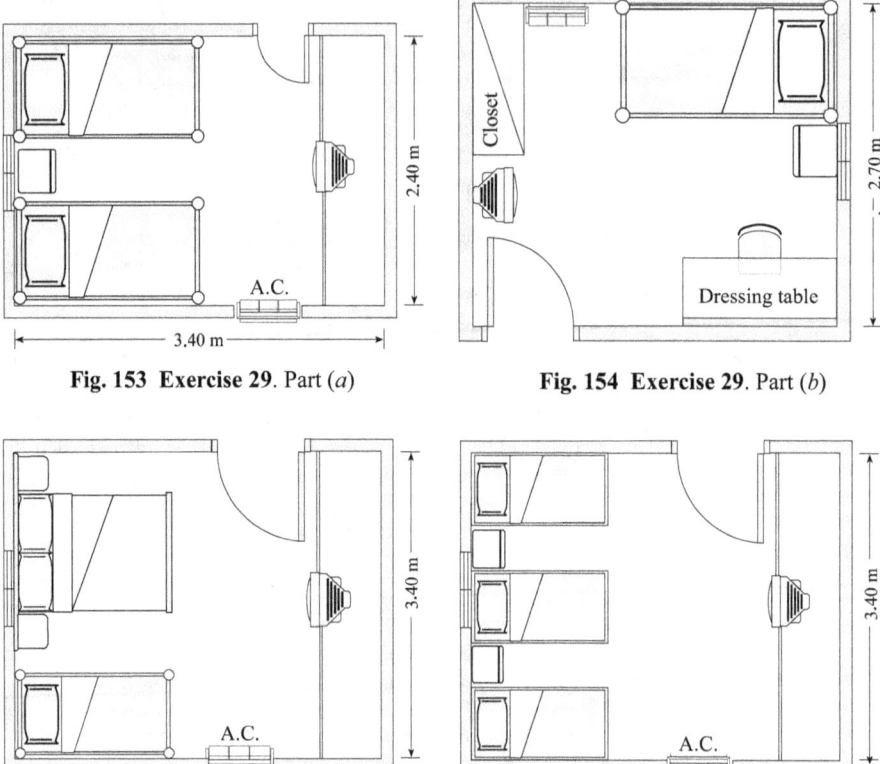

Fig. 153 Exercise 29. Part (*a*)

Fig. 154 Exercise 29. Part (*b*)

Fig. 155 Exercise 29. Part (*c*)

Fig. 156 Exercise 29. Part (*d*)

30. In the floor plans on the following pages, **Figures 157** to **160**, locate, according to what has been studied in this book, the appropriate outlets, lights and switches. (Measurements, in meters, are approximate.)

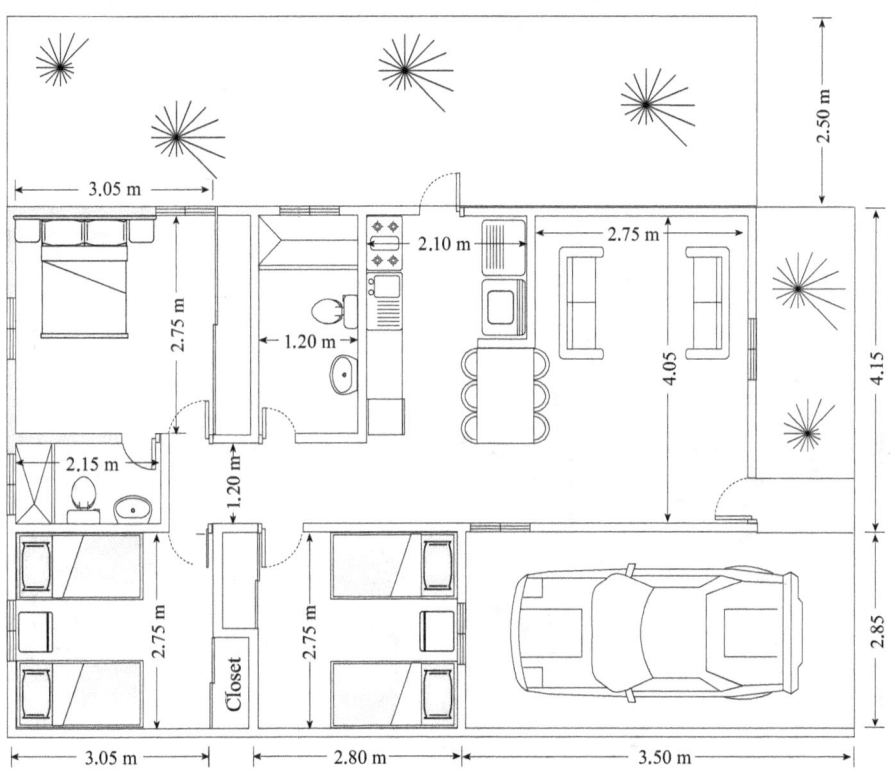

Fig. 157 Exercise 30.

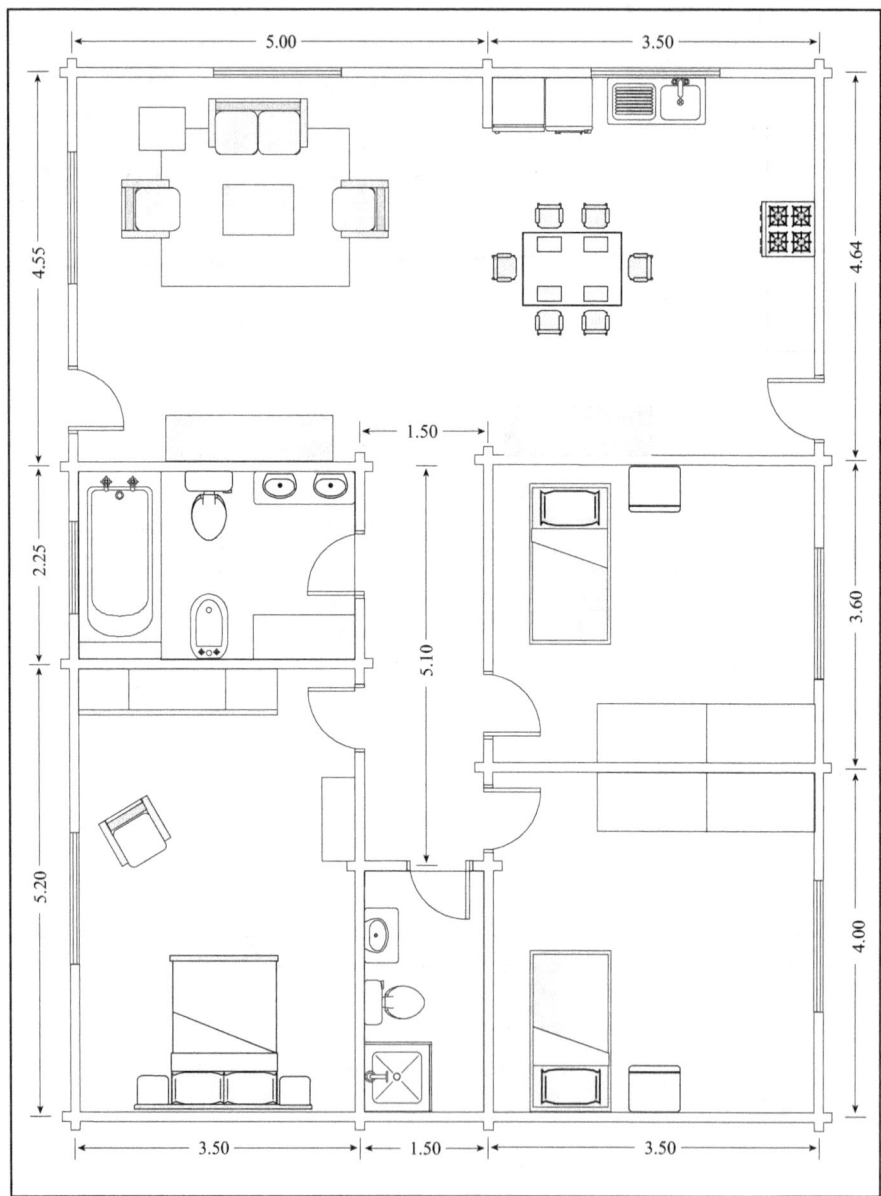

Fig. 158 Exercise 29.

EXTERNAL PATIO

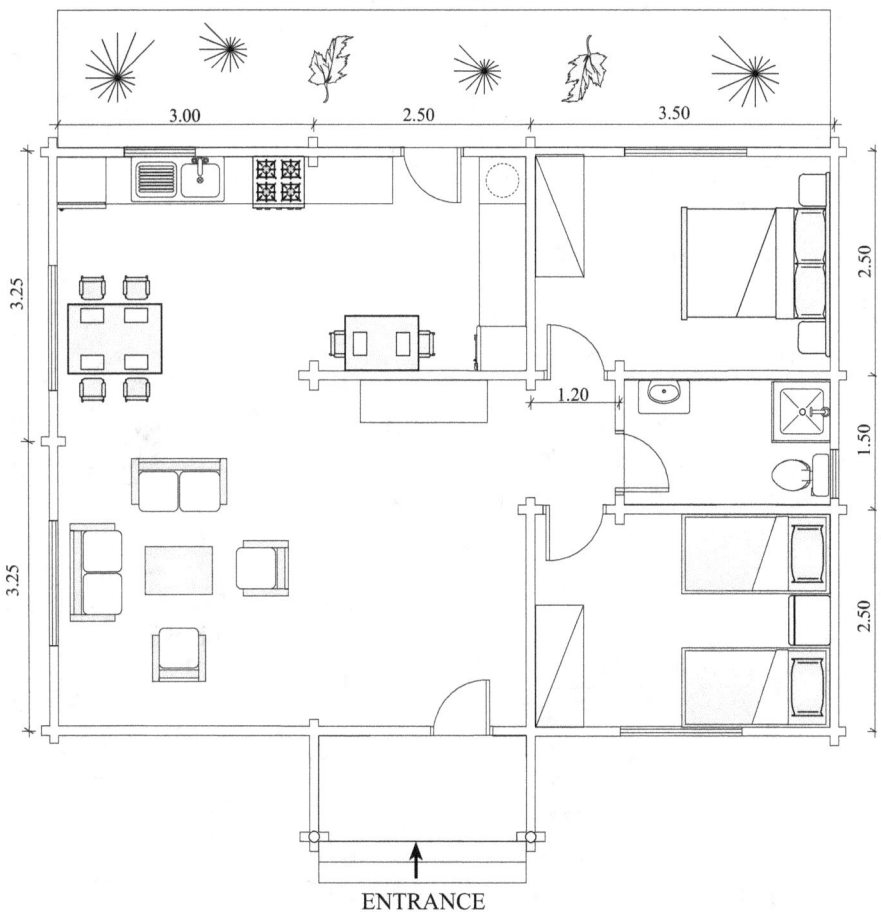

ENTRANCE

Fig. 159 Exercise 29. Part (*c*).

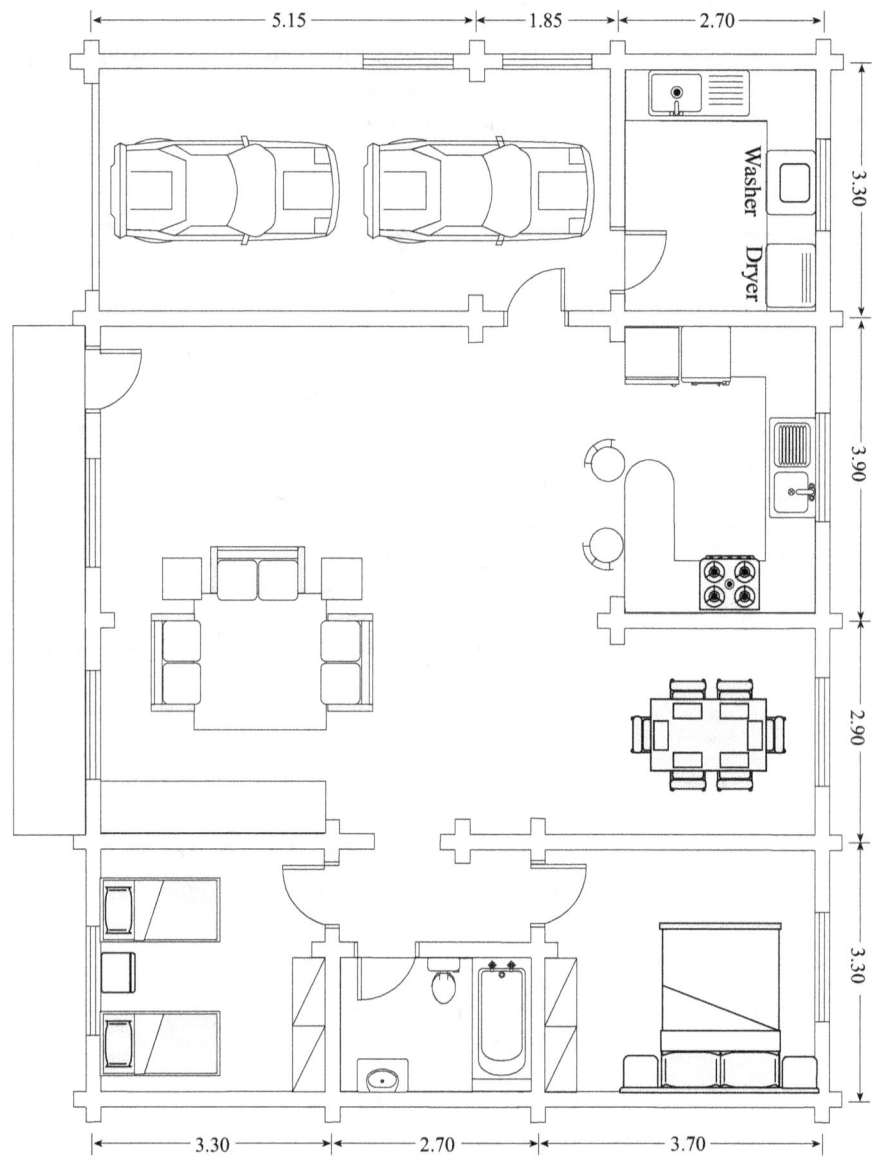

Fig. 160 Exercise 29. Part (*d*).

www.ingramcontent.com/pod-product-compliance
Lightning Source LLC
Chambersburg PA
CBHW081517220526
45467CB00010B/2958